T0398589

BASICS OF
MOLECULAR
RECOGNITION

BASICS OF MOLECULAR RECOGNITION

Dipankar Chatterji

Molecular Biophysics Unit,
Indian Institute of Science,
Bangalore, India

CRC Press
Taylor & Francis Group
Boca Raton London New York

CRC Press is an imprint of the
Taylor & Francis Group, an **informa** business

CRC Press
Taylor & Francis Group
6000 Broken Sound Parkway NW, Suite 300
Boca Raton, FL 33487-2742

Printed on acid-free paper
Version Date: 20151214

International Standard Book Number-13: 978-1-4822-1968-5 (Hardback)

Library of Congress Cataloging-in-Publication Data

Names: Chatterji, Dipankar, author.
Title: Basics of molecular recognition / Dipankar Chatterji.
Description: Boca Raton, FL : CRC Press, Taylor & Francis Group, [2016] |
"2016 | Includes bibliographical references and index.
Identifiers: LCCN 2015047327| ISBN 9781482219685 (hardcover ; alk. paper) |
ISBN 1482219689 (hardcover ; alk. paper)
Subjects: LCSH: Molecular recognition. | Monomers. | Macromolecules.
Classification: LCC QP517.M67 C43 2016 | DDC 547/.7--dc23
LC record available at http://lccn.loc.gov/2015047327

Visit the Taylor & Francis Web site at
http://www.taylorandfrancis.com

and the CRC Press Web site at
http://www.crcpress.com

To all my students

Contents

Preface

THE FUNDAMENTAL ISSUES IN BIOLOGY ARE SELF-REPLICATION, information processing, and turnover of metabolic products to generate energy for sustenance. All these should happen at normal temperature and pressure, thus imposing a great demand on the system. The accuracy with which the system ought to function is always at a premium. In the majority of cases in biology, the accuracy of any process is dictated by the paradigm of molecular recognition, which relies on two basic tenets: two molecules must recognize each other, and there should be three-dimensional complementarities between their surfaces. On the other hand, the molecules need to be macromolecules in nature, which is possible by polymerizing several monomers of same or different types. Such macromolecules generate various shapes and sizes, which ultimately leads to degeneracy in the recognition pattern.

A cursory look at the polymers made with different units and possessing no biological function would indicate that they are not folded according to any defined rules and thus the shapes are not designed specifically for molecular recognition. There are, however, exceptions that will be discussed at the end of this book.

This book attempts to enlist certain principles of molecular recognition. One would hopefully appreciate the enormous efforts of thousands of researches that have been taking place over the past several decades to bring this important subject to a point of admiration.

Acknowledgments

THIS BOOK IS THE OUTCOME OF TWO COURSES I TAUGHT PhD students over the last three decades and undergraduates over the last five years. Molecular recognition is a vast area encompassing every aspect of biology. To bring out the essence of this recognition process is a daunting task, which can be achieved by focusing on specific areas. I relied mostly on DNA-based recognition process, as all aspects of non-covalent recognition are observed here. Sequence specific DNA-protein recognition has always fascinated me. Searching for a specific sequence over DNA by a protein and stabilization of the complex through various interactions is a paradigm of macromolecular recognition process. In addition, sugar-protein recognition, RNA-protein recognition, DNA-RNA recognition are all integral parts of non-covalent macromolecular recognition.

The book begins with the types of bonds that participate in recognition and the functional groups that are capable of forming these bonds. The next topic is how smaller molecules take part in selecting their partners in the overall recognition scheme. Here examples of specific recognition patterns involving molecules other than nucleic acids have been taken. Macromolecular recognition, which is at the heart of the central dogma of molecular biology, forms a large part of this book. A chapter on the selective discussion of various methods that can be used to study molecular recognition has been introduced. The last chapter discusses how molecules without biological function can be arrayed or folded following certain rules and the nature of interactions among them.

In developing this book I was helped immensely by Dr. Sushma Krishnan, a member of my research group, who has also written certain sections at the end of Chapter 3. The help provided by the publishers in way of corrections, suggestions, and helping me keep up the timeline is gratefully acknowledged. Ms. Aastha Sharma, without whose help this book would not have been possible, needs special mention. I also acknowledge the help provided by Ms. Mega Prakash in correcting and editing the book and Ms. Naina and her team for taking care of the artwork. I gratefully acknowledge a discussion with Dr. S. Ramakrishnan. I also put on record here my deep sense of gratitude to different publishers who readily allowed me to reproduce some texts and figures cited at the end as suggested readings. It was possible to finish the indexing of the book on time with the help of Mr. Chandramohan, Indian Academy of Sciences.

Finally, I hope that the student community will benefit from this book.

Author

Dipankar Chatterji, professor, Molecular Biophysics Unit, is at the Indian Institute of Science, Bangalore. He completed undergraduate and postgraduate studies in chemistry at Jadavpur University, Calcutta. Later he earned a PhD at the Indian Institute of Science, Bangalore. His dissertation was titled "Interaction of metal ions with nucleic acids, nucleosides, and nucleotides." Professor Chatterji taught at the University of Hyderabad and also worked on the "Structure-function relationship in *E.coli* ribosome." He then moved to the United States for postdoctoral work at Ken Wu's laboratory at Albert Einstein Medical College, New York, and the State University of New York, Stony Brook, where he worked on *E.coli* RNA polymerase and the role of intrinsic metal ions therein. He started his own laboratory at the Centre for Cellular and Molecular Biology (CCMB), Hyderabad in 1983 and established a major initiative on *E.coli* transcription system. In 1999 he was invited to join the Molecular Biophysics Unit as a professor, where his work on transcription expanded further to mycobacterial transcription machinery. His major area of research is the regulation of gene expression in bacteria under stress, and he has focused on a few important pathways, such as stringent response, quorum sensing, etc. Several students have graduated from his laboratory, and his work has many prestigious recognitions. Professor Chatterji was the president of the Indian Academy of Sciences from 2013 to 2015.

CHAPTER 1

Features of Interacting Monomers with Different Functionalities

What Drives the Binding?

Molecules do not act if they are not bound.

PAUL EHRLICH

The process of how two or more molecules interact with each other is called *molecular recognition*. Molecular recognition is the most important operating language in biology, and it is primarily noncovalent in nature. The term *molecular recognition* was coined, in the 1970s, by Jean-Marie Lehn—a chemist who won the Nobel Prize in 1987.

Molecular recognition plays an important role in chemistry. It forms the basis of highly accurate recognition, reaction, transport, regulation, and other activities in a system for numerous biological processes. Enzymatic reactions, antigen–antibody association, cellular recognition, translation and transcription of the genetic code and several others are based on the recognition patterns of molecules and their interactions.

Molecular recognition is a multistep process in which the molecules or structures recognize each other mainly on the basis of complementarity.

According to Lehn, *molecular recognition* is defined as a process involving both binding and selection of substrates. In the process of binding, the interaction of molecules is based on the following: (1) functional groups, (2) configuration and chirality, (3) size and shape, (4) various interactive bonds, and (5) positioning of the molecules.

In this chapter, we study how small molecules or ligands bind to larger molecules based on the aforementioned properties. First, we see how the presence of different functional groups in monomers plays an important role in small molecular recognition.

1.1 FUNCTIONAL GROUPS IN RECOGNITION BETWEEN SMALL MOLECULES

The functional groups in monomers are independent functional units consisting of atoms. For example, in methane, the carbon atom is tetrahedral and is sp^3 hybridized. Therefore, recognition is possible from any direction. However, if the surrounding atoms are different, the recognition pattern of the molecule will vary.

There can be several types of monomers and they could possess distinctive features; some widely used monomers are described in the following (Figure 1.1).

1.1.1 Functional Groups and Recognition Pattern

It is essential to understand that functional groups are important for molecules to recognize each other. Therefore, this section explains the manner in which these functional groups enable molecules to recognize a partner.

The functional group of alcohol is the −OH group, whose recognition properties will depend on the pair of electrons on oxygen atom or the hydrogen atom bound to it. As the hydrogen is bound to an electronegative centre like oxygen, it can participate in hydrogen bond formation. The recurrent hydrogen bonding in molecular recognition is described in this book.

Similarly, the carbonyl −C=O groups in aldehyde or ketone are polarized to create a partial positive charge on carbon and a partial negative charge on oxygen. This, in turn, forms a hydrogen bond during the recognition. The electropositive carbon center, on the other hand, participates in the charge-based noncovalent interaction with a partner.

Amines form hydrogen bonds with the alcoholic −OH group, and the lone pair of electrons on nitrogen can be donated to an electropositive center. Amides possess the properties of both carbonyls and amines.

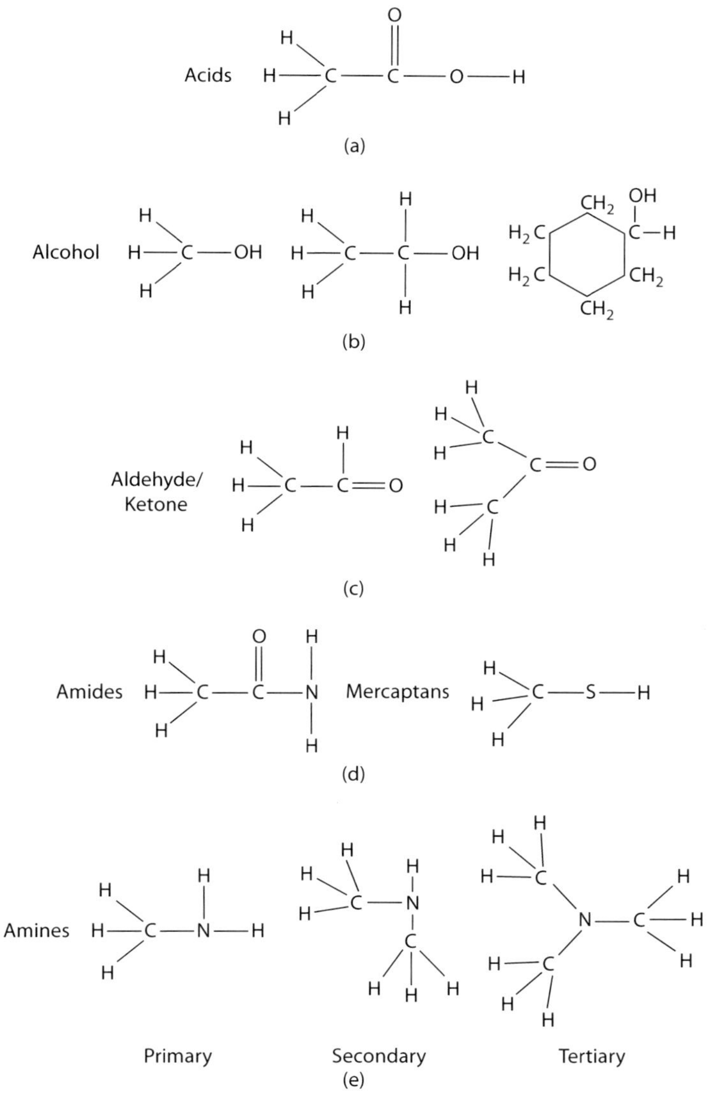

FIGURE 1.1 (a–e) Monomers and functional groups.

The participation of amides in biology is very important and will be seen in Section 1.2.3, which describes the chemistry of peptide bonds.

Mercaptans exhibit similar reaction as alcohols in that oxygens are replaced with sulfurs. They play an indirect role in molecular recognition and involve proteins in the following manner. The native conformation in proteins where the three-dimensional (3D) structures are stabilized by covalent disulfide bonds is altered in the presence of mercaptans due to the reduction of disulfides. This results in different 3D conformation and therefore the protein will have a different recognition pattern.

In natural systems, there are a few fundamental units that are monomers of biological macromolecules and they comprise functional groups such as those mentioned earlier in this section. Sugars in carbohydrates, nucleic acid bases in DNA or RNA, and amino acids and peptides in proteins are a few such examples. Thus, the recognition principle of these molecules is governed by the properties of the different functional groups comprised.

1.2 MOLECULES WITH DIFFERENT FUNCTIONALITIES

1.2.1 Sugars

Sugars exist in two forms: linear and cyclic. The cyclic sugars are the predominant form and the reactivity, especially that of the carbonyl group, is much less due to a reduction of the carbonyls during cyclization. Some important parameters in molecular recognition involve sugars. First of all, sugars have multiple conformations, as each carbon is tetrahedral in nature. The cyclic sugar can exist in two forms: chair and boat. A sugar molecule can also be linked to another molecule through covalent linkage, and this link can take place through different carbon atoms. A variety of possibilities thus exist to generate surfaces of different shapes and sizes by changing the conformation of sugars and the linkage pattern, which are exploited in molecular recognition. A few representative structures of cyclic sugar molecules are shown in Figure 1.2.

Section 2.4 describes proteins with bound sugars, known as glycoproteins, which play important roles in cell surface and regulation, and give rise to variations in protein structure. This specific recognition between sugars and proteins is also important in controlling many functions within cells.

1.2.2 Nucleosides/Nucleotides/Purines/Pyrimidines

Nucleotides are the basic units of nucleic acids. A nucleotide consists of three distinct chemical groups: (1) a nitrogen-rich base [cytosine (C), guanine (G), adenine (A), thymine (T) in DNA, or uracil (U) instead of T in RNA], (2) a 5-carbon sugar (ribose or deoxyribose), and (3) phosphate.

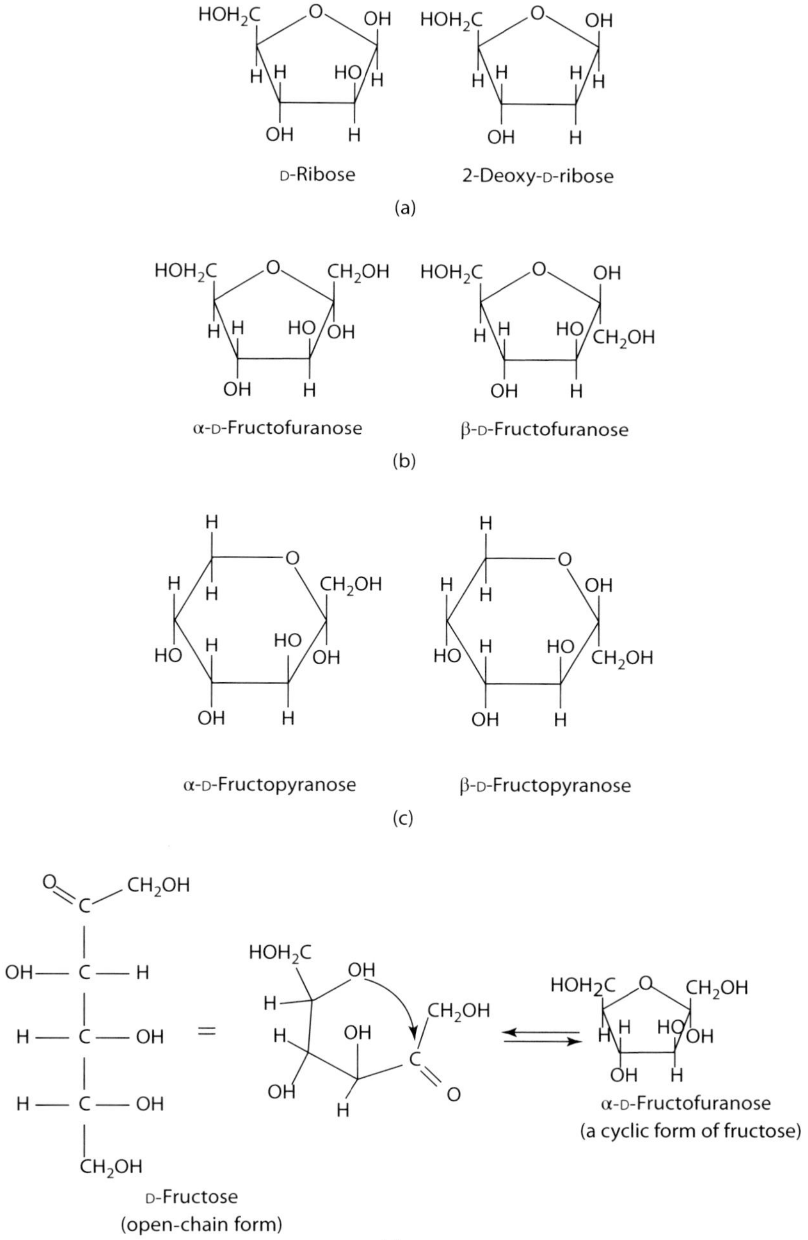

FIGURE 1.2. (a–f) Structures of cyclic sugar molecules.

(*Continued*)

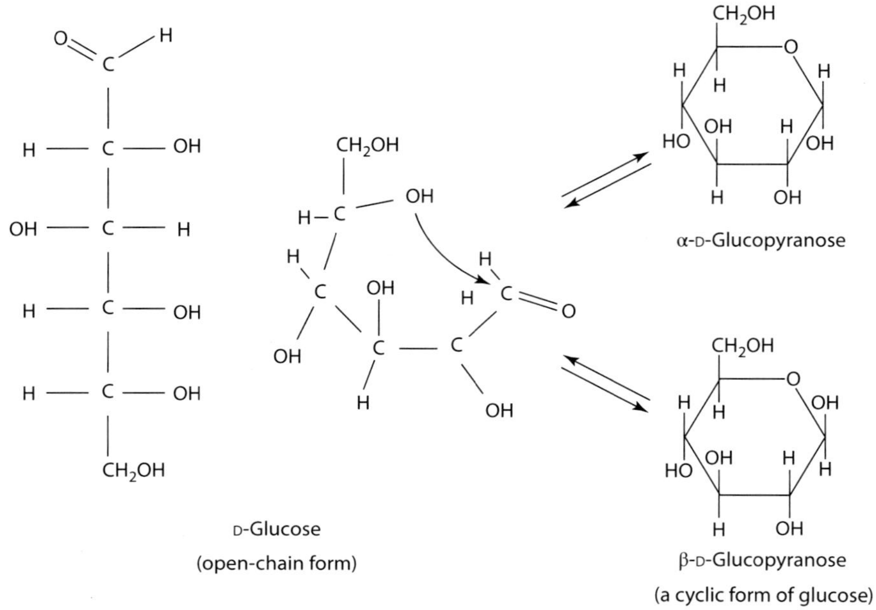

D-Glucose
(open-chain form)

α-D-Glucopyranose

β-D-Glucopyranose
(a cyclic form of glucose)

(e)

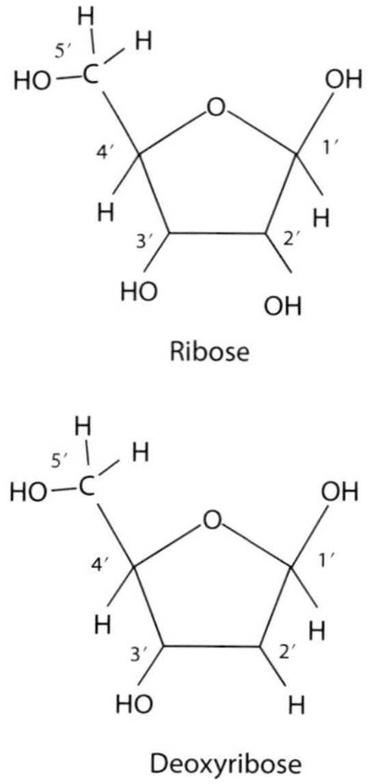

Ribose

Deoxyribose

(f)

FIGURE 1.2 (*Continued*)   (a–f) Structures of cyclic sugar molecules.

Purine and pyrimidine are heterocyclic, nitrogen-rich aromatic bases. In purine, a six-member ring is attached to a five-member ring; however, in pyrimidine, there is only a six-member ring. When a ribose or deoxyribose sugar is attached to one of the nitrogens of the heterocyclic ring, it is called a nucleoside. When a phosphate is attached to the nucleoside, it becomes a nucleotide (Figures 1.3 and 3.3). Interestingly, nucleic acid bases have many nitrogen or oxygen atoms throughout the ring, and these electronegative centers have tremendous potential to act as hydrogen bond donors or acceptors during recognition. The recognition occurs either with another nucleic acid base or with a suitable amino acid in a protein molecule. The aromatic nature of the bases gives rise to the planarity of the structure as well as *pi*-electron stacking with another aromatic group. Both hydrogen bonding and aromatic stacking interactions are important for macromolecular recognition. This theme will appear again in Section 3.3.3, where we will discuss macromolecular recognition involving DNA and proteins.

A few structures shown in Figure 1.3 are nucleic acid bases, and they recognize each other through noncovalent interactions such as hydrogen bonds.

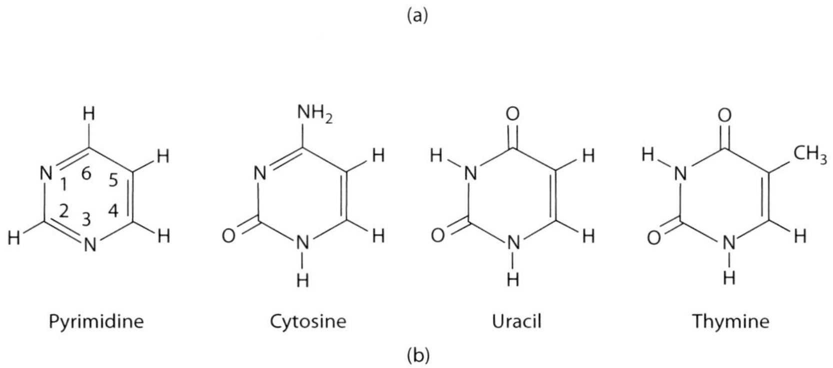

Purine   Adenine   Guanine

(a)

Pyrimidine   Cytosine   Uracil   Thymine

(b)

FIGURE 1.3   (a, b) Structure of nucleic acid bases.

1.2.3   Amino Acids

Known as the building blocks of life, amino acids are the basic units of proteins. They can be acidic, basic, neutral, hydrophilic, or hydrophobic. Due to the nature of their side group (encircled), their properties can vary (Figure 1.4). A peptide unit is formed by a condensation reaction between two amino acids in which a water molecule is eliminated and an amide is formed; this involves a carbonyl group of one amino acid and an amino group of the other (Figure 1.5). The very nature of the amido group results

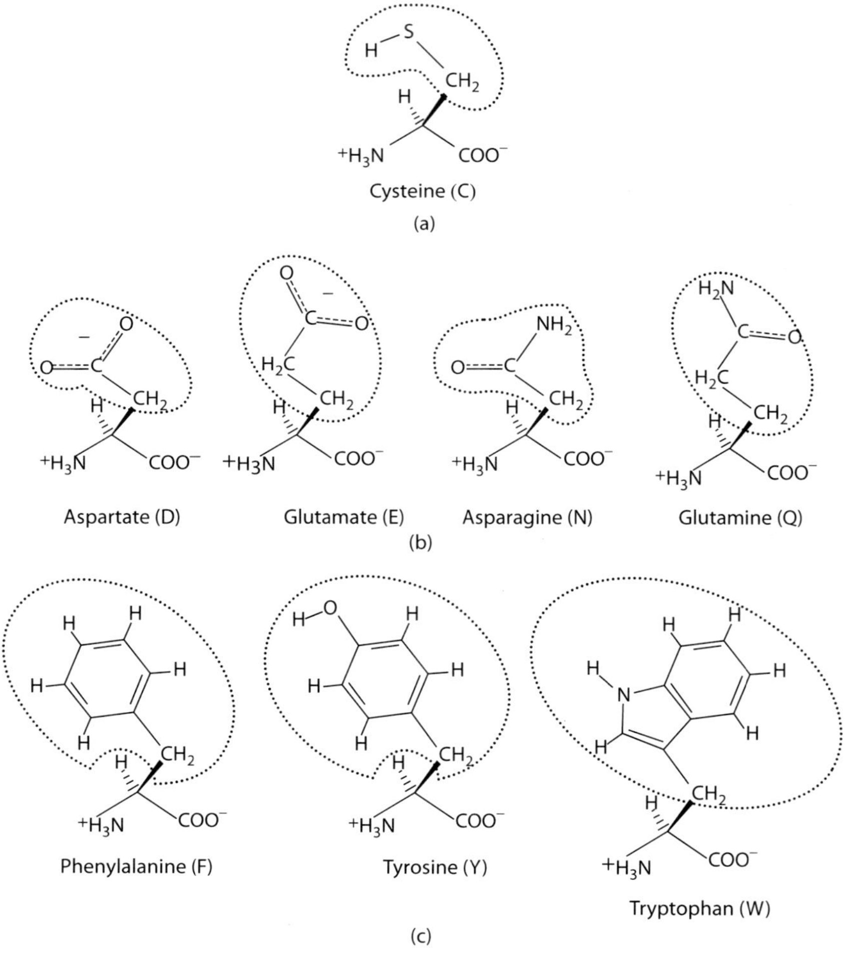

Cysteine (C)

(a)

Aspartate (D)   Glutamate (E)   Asparagine (N)   Glutamine (Q)

(b)

Phenylalanine (F)   Tyrosine (Y)   Tryptophan (W)

(c)

FIGURE 1.4   (a–i) Structure of amino acids.

(*Continued*)

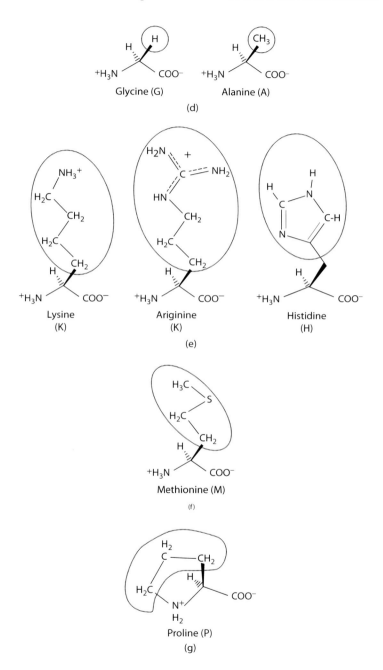

FIGURE 1.4 (*Continued*)   (a–i) Structure of amino acids.

(*Continued*)

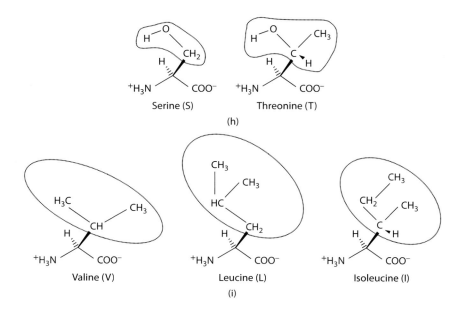

FIGURE 1.4 (*Continued*) (a–i) Structure of amino acids. Single amino acid codes are in bracket.

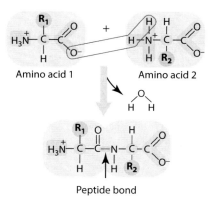

Peptide bond

FIGURE 1.5 Peptide bond formations. (Adapted from Pearson Education, Inc., www.daviddarling.info/encyclopedia/P/peptide bond.html.)

in a 1,2-hydrogen shift (tautomerization), and the peptide bond between C and N becomes either a partial double bond or planar in nature. This feature of the partial double bond as well as the different side group on the carbon atoms attached to the carbonyl or amino group generates a plethora of structural alternatives, which are capable of hydrophilic, hydrophobic, and charged interactions with another partner. Different amino acids could exhibit recognition among themselves or with nucleic acid bases or sugars.

1.3 CONFIGURATION AND CHIRALITY

The recognition of different monomers largely depends on their shape, size, and charge. The water-repelling tendency of monomers also leads to recognition through hydrophobic interactions. Here, the molecules, as such, may not have an affinity toward each other; rather, their dislike for solvent water brings them together in an aqueous medium.

In this section, two important parameters for monomer-specific recognitions are described. They are configuration and chirality.

Configuration is the relative arrangement of atoms that are in space around a central atom. Here, the change in arrangements means bond breaking and bond remaking. It is easy to understand that different positions of the functional groups around the central atom will give rise to varied noncovalent recognition in space with various geometric alignments giving rise to a battery of products.

When the mirror image of a molecule is not superimposable or another molecule with the same chemical composition exists but its mirror image is not superimposable, the molecule is said to be chiral. This is due to an asymmetric center in the molecule. By definition, both molecules have identical properties in all respects except that they interact with plane-polarized light in a different manner. They are also known as *optical isomers*. At times, two isomers are used as two different drugs and, thus, chiral purity in the drug industry is very important. This happens due to different recognition patterns with substrates or enzymes in 3D space.

1.4 LOCK-AND-KEY AND INDUCED FIT MODEL

The recognition of different monomers largely depends on the shape or complementarity shared between them. Emil Fischer (Box 1.1) won the Nobel Prize in 1902 for his lock-and-key model, which he had proposed in 1890. The aim of this model was to explain how enzymes play the role of catalysts in chemical reactions by forming enzyme–substrate complexes. According to Fischer's model, a substrate should have a matching (complementary) shape to fit into an active site of an enzyme (Figure 1.6). The active site is the place where the substrate fits into the pocket of an enzyme. However, it was later proved that enzymes are not rigid in their complementarities. The enzyme and the active site can be modified by the substrate or

BOX 1.1 EMIL FISCHER

Emil Hermann Fischer, popularly known as Emil Fischer, won the 1902 Nobel Prize in Chemistry for his work on sugars, purines, and proteins. Fischer also developed the Fischer projection model, a model that symbolically draws asymmetric carbon atoms. Because of his work on biological molecules, he is known as the "Father of Biochemistry."

Born in Germany, Fischer obtained his PhD from Strasbourg under the guidance of Adolph von Baeyer, who was also awarded the Nobel Prize, but only after his student Fischer had been awarded the prize five years earlier. Fischer also studied under August Kekule, who proposed the cyclic structure of Benzene. Fischer died in Berlin at the age of 66.

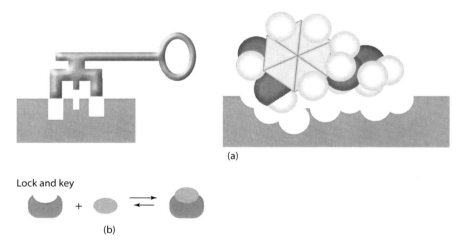

(a)

Lock and key

(b)

FIGURE 1.6   (a, b) Lock-and-key model.

by inducing a "fit" on binding. The induced fit model was postulated by Daniel Koshland in 1958. This model proposed that the initial reaction between an enzyme and a substrate is relatively weak, but this weak interaction induces conformational changes in the enzyme that strengthens the binding (Figure 1.7).

For a successful enzyme–substrate reaction, there are two important requirements. First, the substrate should fit into the enzyme active site; second, the amino-acid side chains of the enzyme need to attract the specific substrate to speed up the reaction.

Induced fit

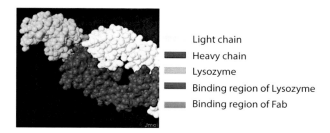

FIGURE 1.7   Induced fit model.

Light chain
Heavy chain
Lysozyme
Binding region of Lysozyme
Binding region of Fab

FIGURE 1.8   **(See color insert.)** Space-filling model of lysozyme binding to its antibody. (http://mcdb-webarchive.mcdb.ucsb.edu/sears/immunology/Antibody-Antigen/hel-fab-f.htm.)

1.4.1   Example of Lock-and-Key Model

The complex of the antigen binding fragment (Fab) of the antibody and the lysozyme is shown in Figure 1.8. The binding surfaces are complementary in shape over a large area.

It is important to note that all the chemical structures shown in Figures 1.1 through 1.4 can take part in noncovalent interactions among themselves or with water to varying degrees (Table 1.1). It should be kept in mind that the protein dynamics play an important role in molecular recognition. It was thought that conformational changes take place due to ligand binding (induced fit) to play the most important role in the recognition process. However the recent study (Chakrabarti et al., 2016) proposed that at least for protein-protein recognition, a conformational state of the protein is pre-selected for ligand binding. These are called "conformationally competent state."

1.5 NONCOVALENT INTERACTIONS

1.5.1 Hydrogen Bond

Hydrogen bonds are formed between two electronegative centers; in these bonds, the H atom is bound to one electronegative center with a covalent bond and shared with the other electronegative center. Hydrogen bonds are longer (0.18 nm) compared with typical covalent bonds involving hydrogen atoms (0.097 nm). When they are linear, they show directionality and

TABLE 1.1   Different Types of Noncovalent Interactions

Type of Noncovalent Interaction	Strength	Description	Example
H-bond	4–4.5 kcal/mol	Sharing hydrogen between two electro-negative centers	Water, alcohol, amines, peptide backbone
Hydrophobic bond	0.7–1.0 kcal/mol	Interaction between two molecules that are water repellent or nonpolar in nature (statistically significant)	Aromatic–aromatic interaction, nucleic acid base recognition, oil in water
Van der Waals' interaction	Weak	Interaction between any two atoms in close proximity	Relevant in protein conformation and folding
Stacking interaction	Weak	*pi–pi* interaction in two stacked aromatic systems due to electron delocalization	Base stacking in DNA double helix, recognition between aromatic amino-acid side chains and nucleic acids
Ionic interaction	Strong between unlike charges and repulsive between like charges	Like charges repel and unlike charges attract	Recognition between charged species, side-chain interactions in proteins

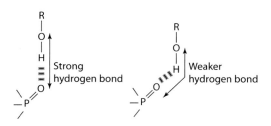

FIGURE 1.9   Bent and linear hydrogen bond. (Adopted from Berg, J.M., Tymoczko, J.L., and Stryer, L., *Biochemistry*, 5th edition, Freeman, New York, 2002.)

are comparatively stronger; however, weak, nonlinear H-bonds are also allowed, as shown in Figure 1.9.

1.5.2 Hydrophobic Interaction

When two molecules that are not soluble in water (water repellent) are placed in a water medium, they come close to each other due to a hydrophobic interaction. These are nonpolar molecules and water molecules form an ordered cage around them. During the interaction, the water molecules around them are released and become free to interact with bulk water. This is a weak interaction but it plays an important role in the process of biological recognition. Usually, a large number of molecules act in a concerted fashion and the hydrophobic interaction becomes a strong phenomenon. Two drops of oil in a bowl of water will coalesce and vice versa due to the hydrophobic interaction.

1.5.3 Van der Waals Interaction

The *van der Waals interaction* is largely defined as an interaction between any two atoms in close proximity. The limit of the interaction is guided by the distance between the interacting partners and is known as the van der Waals distance. This plays a major role in protein–protein interactions as well as in protein folding.

1.5.4 Stacking Interaction

A stacking interaction usually takes place due to delocalization of the *pi* electrons between two aromatic compounds when they are stacked over one another. Aromatic amino acids and nucleic acid bases take part in *pi* stacking. The classical chemical analogy of a stacking interaction is that of the "Ferrocene" molecule, in which two cyclopentadiene ring systems bind the opposite sites of a central iron or any other metal atoms. The aromatic rings sandwich the metal atoms and are stabilized by the aromatic stacking interaction between them.

1.5.5 Ionic Interaction

Two opposite-charged groups are attracted, whereas two same-charged species repel; this has a profound influence on molecular recognition. The nature of the interaction is very strong and is shown in Figure 1.10.

1.5.6 Interaction due to Spatial Match

The spatial arrangements or positioning of atoms or a group of atoms around an asymmetric center may be such that they can fit within the cavity in a protein, which is a large macromolecule. Such recognition processes are mostly observed in the case of sugar–protein interactions. Figure 1.11 shows two subunits of a protein with exact 3D matches that generate a sugar binding site at the junction.

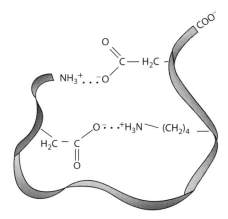

FIGURE 1.10 Salt bridge between arginine and aspartic acid (one with basic and the other with acidic side chains). (Adopted from http://aris.gusc.lv/NutritionBioChem/32Proteins.doc.)

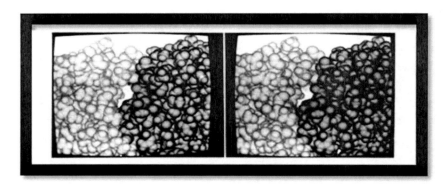

FIGURE 1.11 L-Arabinose binding protein–sugar complex at 2.4 Å resolution. (From Newcomer, M.E. et al., *J Biol Chem.*, 256(24):13213–7, 1981.)

Nonplanarity of the asymmetric center restricts spatial recognition, and monomer–monomer molecular recognition is rare in such cases. However, we find that once the interacting partners are planar there is a wider possibility for the recognition. In the example cited earlier (Figure 1.11), it can be seen that the two proteins interact in such a manner that they create enough space at the crevice for a monomeric sugar to accommodate itself.

CHAPTER 2

Molecular Recognition among Various Monomers

In order to understand how different monomers recognize each other, let us take the example of deoxyribonucleic acid (DNA), which is located in the nucleus of a cell. The DNA is made up of the following bases adenine (A), guanine (G), cytosine (C), and thymine (T), which are held together by two chemical bonds—one is a covalent or phosphodiester bond and the other is a hydrogen bond. Each of these bases is attached to a sugar and a phosphate molecule. Together, a base with a sugar and a phosphate molecule forms the nucleotide, which is a monomer. The hydrogen bond helps the DNA to maintain its stability.

For nucleic acid bases or their variants to be recognized, a minimum of two hydrogen bonds are required. However, three hydrogen bonds are formed only between guanine and cytosine bases. At times, it is also possible for three H-bonds to form between any two given base pairs; however, model building shows that these structures are unstable. With the help of molecular modeling, one can prove that all three H-bonds between two electronegative centers cannot be built in space due to geometric constraints.

In Section 1.2.2, we noted that the most well-known molecular recognition at the monomer level is the formation of DNA base pairs in the double helix (Figure 2.1). Aromatic bases that are attached to two polynucleotide chains that possess hydrophilic sugar–phosphate backbones recognize specific partners through hydrogen bonding. The driving force for such recognition comes from hydrophobic interactions, as the aromatic bases are hydrophobic in nature and turn inside toward each other in water medium. The base-pair formation ultimately stabilizes the double helix. The successive base pairs in the polynucleotide chains that are oriented inside the aqueous medium are lined up and participate in stacking interactions. Therefore, in a DNA double helix, three interactions are in operation: one covalent (phosphodiester), hydrogen bonds (between base pairs) of two strands of the helix, and a stacking interaction between base pairs on the same strand (Figures 3.5 and 3.6, next chapter).

Specific noncovalent recognition among monomers is otherwise extremely rare, although aromatic amino acids are known to identify specific DNA base pairs, especially when they are a part of a larger protein or DNA molecule. Figure 2.2 shows how an amino acid can bind to a base pair without disturbing the hydrogen bonding schemes between base pairs. Such an interaction can be visualized in nucleic acid–protein recognition.

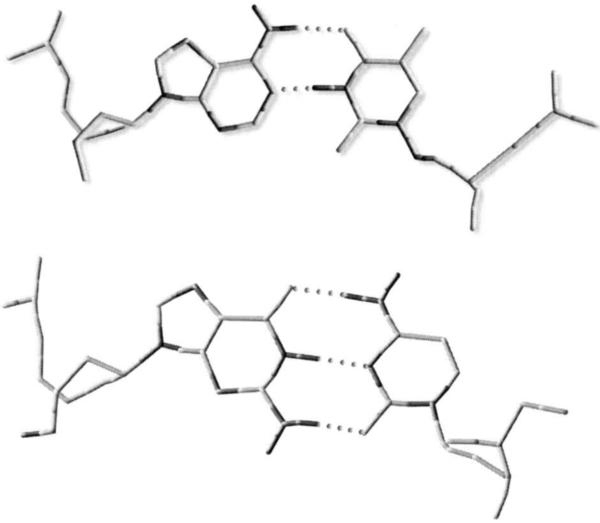

FIGURE 2.1   A–T (top) and G–C (bottom) base pairs. A base pair is held together by hydrogen bonds between the purine base (two fused ring system) and the pyrimidine base (one ring). At the top, there are two hydrogen bonds between A and T; at the bottom, there are three hydrogen bonds between G and C. (Adapted and modified from http://molvisual.chem.ucsb.edu/nucl_struc.html.)

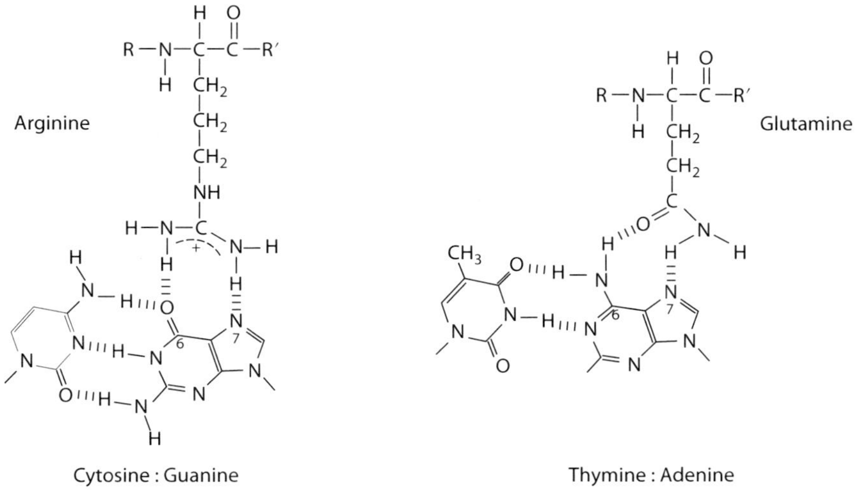

FIGURE 2.2 Binding of amino acid to nucleotide. In this example, glutamine or arginine forms hydrogen bonds with A–T and G–C base pairs, respectively. (Adapted and modified from http://www.bioinfo.org.cn/book/biochemistry/chapt27/sim2.htm.)

It is very difficult to quantitate the stability of a noncovalent interaction between two molecules when they are not a part of a larger network. Even when nucleic acid bases are free in solution, they exhibit a weak interaction. Macromolecular recognition is associated with a decrease in entropy (due to a decrease in disorder); however, the recognition takes place spontaneously due to free energy minimization. In the absence of a macromolecular network, the minimization of free energy is insignificant.

Molecular complementarities play a major role in recognition, as discussed in Section 1.4. In addition, the solvent system can aid in driving the interacting partners toward each other. Conceptually, this is very close to hydrophobic interactions. In Chapter 3, the driven assembly process will be discussed in detail. In Chapter 4, the methods to follow these assemblies will be shown.

If one of the reacting partners is a macromolecule, the specificity in the process of recognition becomes predominant and guides many important biological and chemical processes. Although there are several macromolecular recognitions—such as DNA–protein recognition, antigen–antibody interaction, and RNA–ribosome interaction—in biology, few important monomer–macromolecule recognitions are worth noticing. Here, we consider recognitions such as crown ether complex formation with metal ions, peptide–antibiotic interactions, drug–receptor interactions, and sugar binding to proteins.

2.1 CROWN ETHER–METAL ION RECOGNITION

Crown ethers are cyclic compounds containing ether linkages; they could be termed *oligomers of ethylene oxides*. For example, 18-crown-6 complexes with potassium ions, 15-crown-5 has affinity for sodium ions, and 12-crown-4 has affinity for lithium ions (Figure 2.3). Metal ion–bound crown ether resembles a king's crown, hence the name. The nomenclature is simple: the first number represents the number of atoms in the cycle, and the second number represents the number of oxygen atoms. Crown ethers are generally hydrophobic in nature. However, they can complex with monovalent or divalent cations placed inside the pore of the crown ether. The oxygen atoms coordinate with the metal ions.

2.2 PEPTIDE–ANTIBIOTIC RECOGNITION

The second example of molecular recognition that has biological significance is the peptide–antibiotic interaction.

Antibiotic vancomycin specifically binds to the peptides with the help of the terminal dipeptide sequence D-alanyl-D-alanine in bacteria (Figure 2.4). The specificity of the interaction is manifested with five hydrogen bonds. This way vancomycin inhibits the use of these peptides for the construction

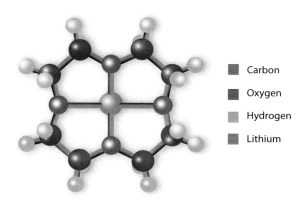

Carbon

Oxygen

Hydrogen

Lithium

FIGURE 2.3   **(See color insert.)** 12-Crown-4 lithium-ion complex. (Adapted and modified from http://www.chem.ucla.edu/harding/crownethers.html.)

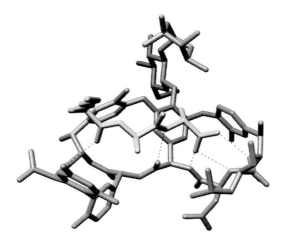

FIGURE 2.4 **(See color insert.)** Crystal structure of L-Lys D-Ala–D-Ala with vancomycin (green represents the antibiotic, and blue represents the tripeptide). (From Knox, J.R., and Pratt, R.F., *Antimicrob Agents Chemother*, 1342–1347, 1990, https://en.wikipedia.org/wiki/Vancomycin.)

of bacterial cell wall. The specificity of the interaction is strong with five hydrogen bonds (shown as dashed lines in the Figure 2.4, compare with GC base pairs) and the peptides are not released for any function.

2.3 DRUG–RECEPTOR INTERACTION

A small drug molecule, essentially organic in nature, needs to bind specifically to a receptor to exert its functions. The most important event in this interaction is the chemistry and the spatial match between the drug and the interaction surface in the receptor (Figure 2.5). Thus, aromatic drugs exhibit good binding with receptors with the help of *pi* electron clouds. However, from the conformational standpoint, they do not offer a great variability due to their planar structure; thus, their potential as drugs is partly compromised. Many drugs with nonplanar substitution increase the selectivity on binding with the receptor. Although most of the receptors are protein molecules, there are two different types of drugs:

- Agonists—stimulate and activate the receptors

- Antagonists—prevent the agonists from stimulating the receptors

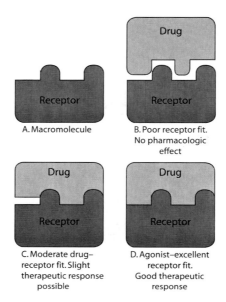

A. Macromolecule

B. Poor receptor fit. No pharmacologic effect

C. Moderate drug– receptor fit. Slight therapeutic response possible

D. Agonist–excellent receptor fit. Good therapeutic response

FIGURE 2.5 Drug–receptor interaction. A perfect match is necessary for strong three-dimensional complementarities in the drug–receptor interaction.

Determination of the affinity between the drug and its receptor presents a formidable challenge. Various methods that aid in designing a drug molecule are available. The nonbonded interaction between the drug and its receptor can be estimated based on the free energy change, which in turn is the sum of the free energy change due to electrostatic, hydrophobic, inductive, and stacking interactions. Computer-based calculations are routinely used to predict the conformational match between the surfaces of the drug and its receptor. Several program packages are available for drug design. However, it is important to recognize that the strength of binding is not the only determining factor for drug design. The nonbonded interaction should be such that the drug can also be dissociated from the receptor with a small trigger, at an appropriate site. The conformational flexibility plays a major role in this direction.

The drug–receptor interaction involves (1) redistribution of charge within the drug or the receptor, referred to as polarization and (2) redistribution of charge between the drug and the receptor, referred to as charge transfer. The drug–receptor interaction generates attractive power between them.

The effectiveness of a drug is determined by its ability to change the surface of the interface due to the redistribution of charges.

2.4 SUGAR–PROTEIN RECOGNITION

Sugars or carbohydrates can bind proteins or lipids, thus generating diverse biological functions, starting from the regulation of gene expression to cell–cell communication. Polymerization of sugars is carried out enzymatically; however, there is no defined template or code for them. In this aspect, they differ from nucleic acid or protein synthesis. Sugars are different in another aspect in that they do not fold and, thus, occupy large surfaces. This property is cleverly utilized in cell–cell recognition. Sugar that binds to proteins at the cell surface has the potential for molecular recognition, as it utilizes a large surface area. Sugars at the cell surface can interact with other cells through the sugars that decorate the proteins. There are a large group of proteins called lectins; these specifically bind sugars with a high affinity. In this section, we emphasize three major recognition processes involving sugars: (1) as inducers for their own metabolism to produce energy, (2) for cell–cell communication, and (3) lectin–carbohydrate interactions.

Sugars are the major energy source for biological species and they are often utilized in a preferential manner. An entire network of the genetic regulatory system is activated in the presence of the sugar (inducer) so that the protein products from the genes are efficiently utilized to metabolize a particular sugar. In the absence of the inducer, the whole genetic system is either deactivated or repressed.

A protein molecule carries the message embedded in DNA base sequence through the processes of transcription and translation. The molecular recognitions that play an important role in these two processes are dealt with in Section 3.5. However, an enzyme that is required for metabolizing a particular sugar will be synthesized only in the presence of the sugar substrate. This beautifully controlled circuit operates on the basis of a defined recognition principle. A protein molecule known as a *repressor* is synthesized in the absence of the sugar, which eventually binds DNA (operator region) and prevents the movement of RNA polymerase. Thus, the transcription of the genes that are necessary to metabolize the sugar is kept under control. The process of derepression (activation) involves the binding of the repressor molecule with the sugar. As a consequence, the repressor does not bind the DNA sequence downstream of the RNA polymerase binding site (promoter region). This removal of the roadblock enables the RNA polymerase to roll on, thus facilitating RNA synthesis. In the absence of the inducer, the synthesis of proteins, which metabolize the inducer, becomes redundant, as the free repressor binds DNA and the roadblock is reestablished (Figure 2.6).

When we try to understand the principle of molecular recognition, sugar-induced activation and the repression story serve as a unique paradigm. Here, the entire control circuit aims at utilizing the substrate sugar as per the necessity. If the cell does not need the sugar for its survival, then at any given point of time during growth, the circuit will not be operational. This process involves protein–protein recognition, sugar–protein recognition, and DNA–protein recognition through different surfaces in the protein molecule. A short-circuit in the control circuit can be imposed (mutated), which results in the uncontrolled synthesis of protein. This is known as *constitutive expression*. Most of these mutations take place at the surface of the recognition.

Often, the repressor molecules are multimeric proteins and the interaction of sugars with these proteins is cooperative (allosteric) in nature. Cooperativity ensures the specificity of this interaction and the tightness of the binding beyond a certain concentration. As expected, the DNA sequence (operator) on which the repressor binds also has a twofold symmetry (palindromic) that matches the symmetric repressor molecule. In fact, one would find that macromolecular recognition in biology is guided by matching symmetry, which generates extra stability. On the other hand, small molecules such as sugars bind repressor proteins following the cooperative phenomenon to regulate association–dissociation more stringently beyond a certain concentration range.

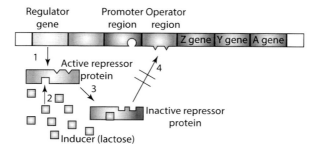

FIGURE 2.6 Regulation of gene expression induced by the sugar molecule. The active repressor will bind the operator region; however, in the presence of sugar, the repressor will bind the sugar.

Several decades earlier, Jacques Monod, a French scientist, noticed that when a mixture of sugars is administered to a growing bacterial population, one sugar is utilized at a time; that is, the bacteria use the second sugar only if the first sugar is exhausted. This phenomenon was named the *diauxic growth pattern*; according to this phenomenon, the specific sugar–repressor interaction controls the regulated use of the sugar, as described earlier. The organism takes time to get adjusted to the new sugar utilization system, and this process of adjustment is known as *adaptation* (Box 2.1).

BOX 2.1   JACOB AND MONOD

Utilization of sugar lactose by coordinated regulation of a set of genes is a paradigm in prokaryotic gene regulation. This gave birth to the concept of the **lac** operon. Monod (together with Jacob) came up with a model that describes how different proteins are synthesized in a controlled fashion, the sole purpose of which is the use of the substrate lactose. They designed the concept of the repressor protein, which binds the operator sequence in DNA and prevents the transcription of adjacent genes. Monod worked on this problem from his PhD days and even as he underwent a very difficult time in Paris during the German occupation. He participated in the French resistance movement and is considered a leading intellectual of France. Monod and Jacob obtained the Nobel Prize in 1965 for their work. Monod also wrote a famous book, *Chance and Necessity: An Essay on the Natural Philosophy of Modern Biology*, which describes his philosophy "Nature does not have any intention or goal."

François Jacob   Jacques Monod, 1910–1976

2.5 CELL WALL COMPOSITION AND SUGARS

Bacteria are the most abundant forms of life. They use several survival strategies that enable them to live in the habitat that they choose. To cope with environmental changes such as temperature, pH, and free radicals, the bacterial cell membrane act as a barrier that helps the bacteria to adapt. Bacterial cell membranes are composed of lipids, sugars, and proteins that form a network, similar to forming a tight envelope, which enables processes such as nutrient transportation to be performed only when assisted by other proteins. The nature of these proteins is largely hydrophobic, and they recognize sugars and lipids through noncovalent interactions. However, covalent modifications of proteins are also known for sugar–protein and lipid–protein complexes. There are receptor molecules on membranes that bind to specific small sugars or nucleotides and transport them from outside to inside and from inside to outside as per the cellular need.

These receptors are proteins in nature with hydrophobic domains spanning across membranes. The interaction between membrane lipids and receptors is hydrophobic in nature, whereas a receptor–ligand interaction can be either hydrophilic or hydrophobic depending on the environment in which the recognition takes place.

2.6 CELL–CELL COMMUNICATION AND SMALL MOLECULE–RECEPTOR RECOGNITION

There are small nucleotides that are known as second messengers. These messengers participate in the signal transduction cascade in biological systems. Organisms have the ability to sense and respond to environmental fluctuations in order to ensure their survival by regulating cellular metabolism, which is known as *signal transduction*. The signal transduction mechanism helps cells detect and amplify the extracellular signals that are received by them and converted into cellular processes. The principle of molecular recognition becomes apparent when the mechanism of signal transduction is revealed. The extracellular signals are unique chemical substances, which may be amino-acid derivatives, small peptides, nucleotides, or proteins. Changes in the environmental conditions will trigger the secretion of small chemicals (ligands or primary messengers) from the inside to outside of the cells. Secondary messengers are intracellular signaling molecules that are generated by cells to modulate cellular changes as per environmental cues. When a ligand or a first messenger binds to a receptor, the second messengers are produced that relay the signals to the effector to carry out a change in the cellular metabolism. Thus, the interaction between the ligand and the receptor (most often proteins) must be very specific, as it serves as a unique example of molecular recognition (Figure 2.7). In the receptor, the conformational change caused by the ligand–receptor interaction often stimulates the enzyme within a cell; this results in producing multiple products or small nucleotides (second messengers), some important ones of which are shown in Figure 2.8.

Thus, when cells experience extracellular signals known as *first messengers*, they respond by generating secondary messengers. This is necessary for cell–cell communication. Bacteria can sense the presence of another bacterium in the neighborhood through the production of small molecules that participate in cell–cell communication.

The targets of second messenger nucleotides are different. They bind to either the effectors, which will regulate cellular functions, or RNA polymerase to regulate specific gene expression. However, their nature

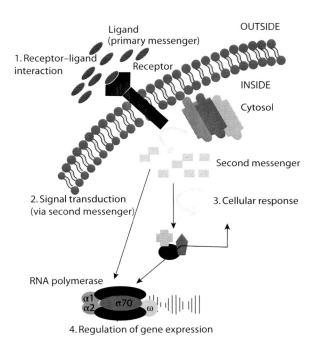

FIGURE 2.7 **(See color insert.)** Schematic representation of a signal transduction pathway in bacteria. A primary signal from outside the cell is transferred inside an effector molecule through a series of recognition processes. (Adapted and modified from Bharati, B.K., and Chatterji, D., *Curr. Sci.*, 105(5), 643, 2013.)

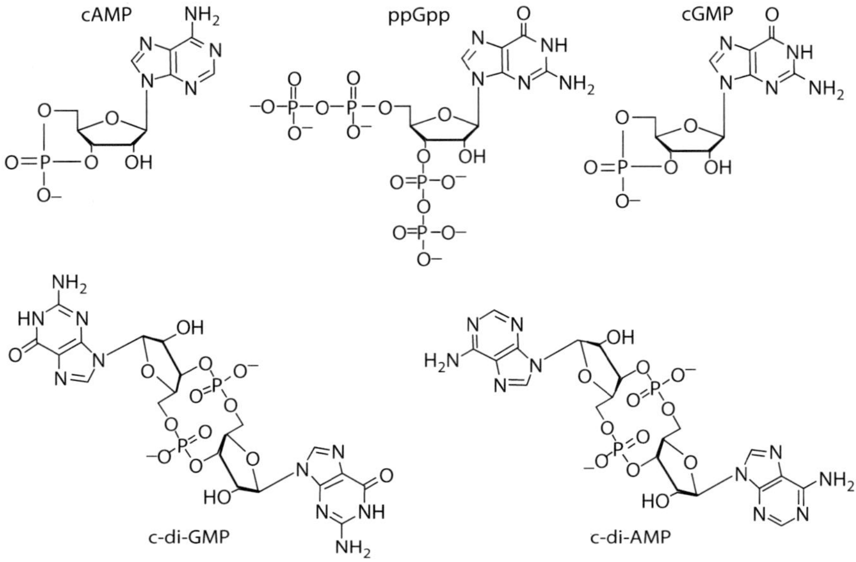

FIGURE 2.8 Intracellular mono- and di-cyclic nucleotide signaling molecules in bacteria. Representative small-molecule second messengers are shown here. Most of them target the transcription system in bacteria. (Adapted from Shanahan, C.A., and Strobel, S.A., *Org. Biomol. Chem.*, 10, 9113–9129, 2012.)

of interaction dictates that the specificity of small molecule–protein recognition is the key for such regulations. Four different classes of second messengers along with their putative targets are shown in Figure 2.9. The study of molecular mechanisms of external signal detection by the bacterial signal transduction system and their regulatory effects on the cellular metabolism is the most important process in the current practice of medicine.

2.7 LECTIN–SUGAR RECOGNITION

In Latin, "lectus" means "to select." Although it was first discovered in plants, lectin is present throughout nature. Conconavalin A is the most common lectin; it has a very strong sugar-binding ability, even though its binding surface is very small. The association between lectins and carbohydrates is multivalent in nature, resulting in specific cell–cell association (Figure 2.10). Lectins are ubiquitous in nature, do not exhibit enzymatic activity, and are not produced by the immune system. In animal cells,

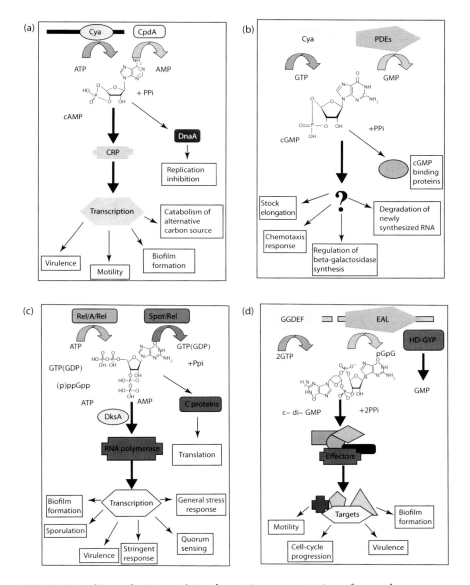

FIGURE 2.9 **(See color insert.)** A schematic representation of second messenger signaling modules in bacteria. Cyclic AMP, cyclic GMP, ppGpp, and c-di-GMP signaling modules are shown in (a)–(d), respectively. (Adapted and modified from Pesavento, C., and Hengge, R., *Curr. Opin. Micrbiol.*, 12(2), 170–176, 2009.)

FIGURE 2.10   (**See color insert.**) Structure of an oligosaccharide bound to a part of the lectin plant. (http://mbu.iisc.ernet.in/~kslab/.)

lectins are involved in cell adhesion and glycoprotein synthesis. In both these cases, specific sugar binding plays an important role. In plants, however, the role of lectins is obscure. It is believed that lectins take part in germination and survival of seeds.

CHAPTER 3

Macromolecular Recognition

3.1 STATIC AND DYNAMIC MOLECULAR RECOGNITION

Molecular recognition could be either static or dynamic. In static molecular recognition, a molecule perfectly binds to a specific binding site, similar to a key fitting seamlessly only into a specific keyhole. This type of specific binding is possible due to a decrease in entropy among reacting partners, which, in turn, is compensated by the removal of water from the interacting surface. The water molecules that are liberated as a result of this process become a part of a large network of bulk water, thus maximizing the entropy. Static recognition has one binding site, is necessarily stoichiometric, and consists of a 1:1 ligand-to-receptor ratio.

Contrary to static molecular recognition, when a molecule binds to multiple binding sites, dynamic molecular recognition takes place. This recognition is governed by multiple binding constants between ligands and receptors.

In each successive step, the binding of one ligand influences the binding of another ligand. This serves as the hallmark of an allosteric interaction, which is the result of conformational change in the receptor due to ligand binding (Figure 3.1).

In dynamic molecular recognition, the binding of an inducer at a particular site on the receptor induces conformational alteration at a distant

Static:

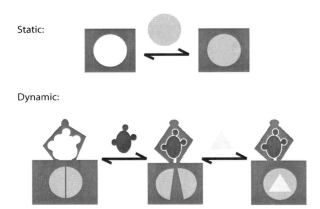

Dynamic:

FIGURE 3.1   **(See color insert.)** Static and dynamic binding

Sigmoid curve of growth of population (S curve)

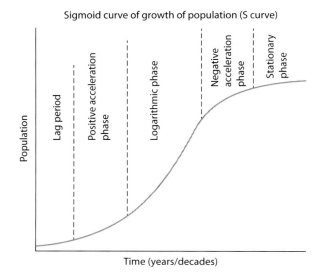

Time (years/decades)

FIGURE 3.2   Sigmoidal growth curve. (Adapted and modified from http://www.tutorvista.com/.)

site, thus making it amenable to recognize a suitable ligand. This kind of interaction is known as an allosteric interaction and is characterized by a sigmoidal binding pattern, as shown in Figure 3.2. The sigmoidal binding pattern is associated either with the growth of any species or with allosteric recognition between a ligand and a receptor, for instance, the oxygenation pattern of hemoglobin. An important phenomenon characterizing molecular recognition is the interdependence of two sites for the purpose of ligand binding.

It is easier to follow dynamic recognition, as it deals with larger macromolecules, compared with static recognition, which involves two single monomers. There are many instances in which macromolecular recognition gives rise to both thermodynamic stability and dynamic molecular recognition. Thus, it is obvious that a polymeric chain, which is biological in nature, generates a specific recognition pocket by folding; this could also influence an allosteric interaction.

3.2 DIFFERENCE BETWEEN CHEMICAL AND BIOLOGICAL MACROMOLECULES

Each structure within biological macromolecules performs a specific function that is governed by folding of the macromolecule along a predetermined path. This structure–function relationship is, therefore, an integral aspect of molecular recognition. Unlike the case of chemical macromolecules, the direction of recognition between two parts of the same molecule or between different molecules gives rise to a proper folding pathway. Anfinsen's experiment on ribonuclease showed that the three-dimensional (3D) native structure of a protein is portrayed through its minimum free energy conformation, which is important for the biological function of the enzyme. Thus, any deviation from the folding or recognition between different parts of the polypeptide chain will render the enzyme inactive.

3.2.1 Ribonuclease

Ribonuclease (RNase), which consists of a group of enzymes, nonspecifically cleaves ribonucleic acid (RNA) molecules. RNA molecules (one of the varieties of nucleic acids) are central to the synthesis of protein molecules that are transcribed from the deoxyribonucleic acid (DNA) base sequence. There is no sequence specificity for most of these enzymes and they can cleave any RNA molecule, most of which are single stranded. Ribonuclease is 124 amino acids long in bovine species; 8 cysteine residues form specific disulfides with each other, generating 4-S-S linkages. These linkages act as strong clips, as the bond is covalent in nature. Anfinsen (Box 3.1) showed that specific positioning of amino acids and S-S linkages is important for the substrate recognition; in this case, an RNA molecule functions as the substrate. A faulty disulfide

BOX 3.1 CHRISTIAN ANFINSEN

Born in 1916, Christian Boehmer Anfinsen Jr. was an American biochemist who studied the relationship between the structure and function in enzymes, a problem encountered in biochemistry. His work on ribonuclease (an enzyme that digests RNA in food) won him the 1972 Nobel Prize in Chemistry that he shared with Stanford Moore and William Howard Stein. On the basis of his studies conducted on ribonuclease, he proposed that the native structure of small globular proteins is determined by the sequence of their amino acids. He proposed that "at least for a small protein the native structure is unique and the thermodynamics of folding guide the protein to its native structure." This hypothesis is well known as "Anfinsen's dogma" or the "thermodynamic hypothesis."

bond leads to scrambled conformations of the enzyme, resulting in the loss of original activity.

In his experiment, Anfinsen demonstrated that the sequence of amino acids and their positions are important for biological functions. However, in chemical polymers, the repeating units are similar, unlike in nucleic acids, where there are 4 different bases and 20 different amino acids for proteins. In chemical macromolecules, the polymerization process is not guided by any specific rule of folding or directionality. Therefore, in such cases, the formation of any recognition domain is limiting. In Chapter 5, we will discuss a few attempts that are being made in this direction.

3.3 BIOLOGICAL MACROMOLECULES

A living cell consists of several macromolecules, each of which performs independent functions. Often these macromolecules interact with each other following the principles of molecular recognition. One may recall having discussed some of these principles in Chapter 1. In this chapter, we will once again discuss the nature of monomers, the polymerization processes, and the manner in which functional sites within the macromolecules are generated; non-covalent recognition and active site conformation. We will also study three different kinds of polymers (nucleic acids, proteins, and sugars) and the process of recognition that they share.

3.3.1 Nucleosides, Nucleotides, and Nucleic Acids

Nucleic acids are the building blocks of DNA or RNA. These bases are covalently linked to cyclic sugars that consist of five carbons called *ribose* or *deoxyribose* at a specific position; this link is known as a *glycosidic bond*. Figure 3.3 shows that the numbering of the members is different for the

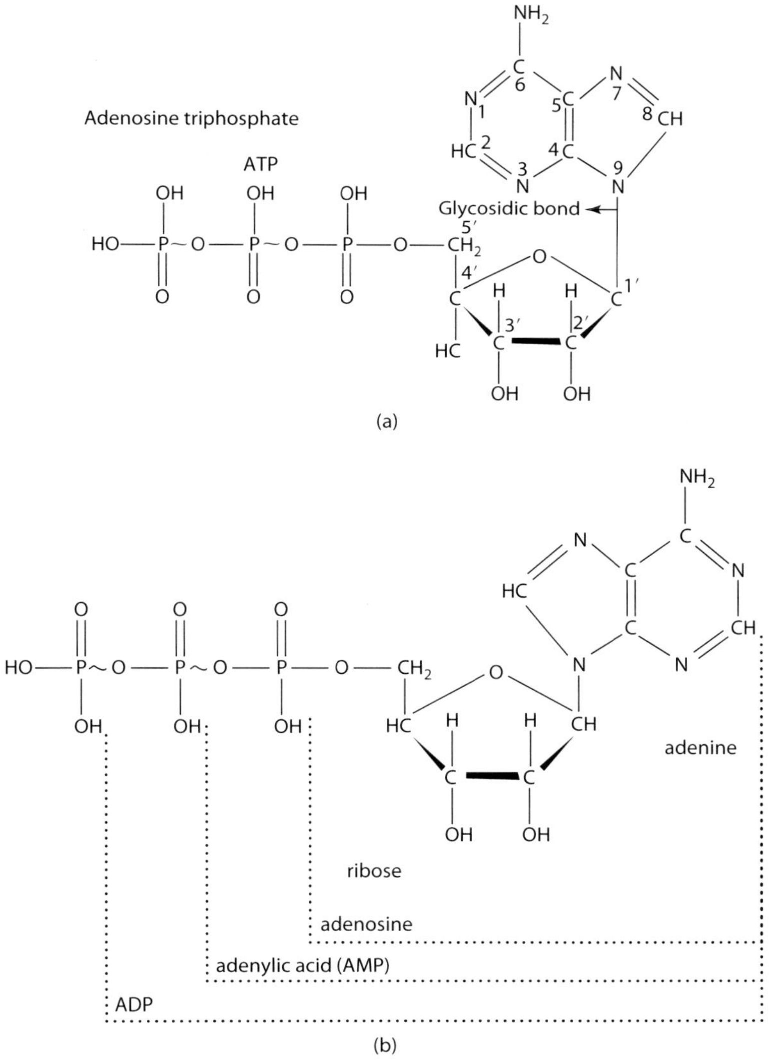

FIGURE 3.3 Chemical structure of ATP. (Adapted and modified from https://en.wikibooks.org.)

heterocyclic base and sugars. The carbon atoms in sugars are denoted with a prime, and are known as *nucleosides*. Similarly, when 5-prime carbons of the sugars are attached covalently with phosphate groups, they are known as *nucleotides*. The phosphates could be mono-, di-, or tri-phosphate, as shown in Figure 3.3.

When a triphosphate is attached to a sugar–base complex, it is known as a nucleoside triphosphate, as shown in Figure 3.3. This structure makes it easy to visualize that the molecule is unstable due to the proximity of the negative charges and coulombic repulsion (curly bonds ~). However, by using the principle of molecular recognition, these molecules can be stabilized with positively charged metal ions. The nucleoside triphosphates play diverse functions, such as being a rich energy source for living systems, in signal transduction, and as various cofactors and substrates.

Two successive nucleotides can be joined by linking the 5-prime phosphate with the 3-prime hydroxyl of the next nucleotide, generating a 5-prime-3-prime-linked dinucleoside monophosphate (Figure 3.4). Several such successive dinucleotide units can be joined together, thus forming a single chain of nucleic acid (Figure 3.5).

A polymeric chain containing several such units will have a different array of bases that are selected out of the four bases that are routinely available for building the nucleic acid structure. Similarly, a second polynucleotide chain running in the opposite direction can be generated. If the

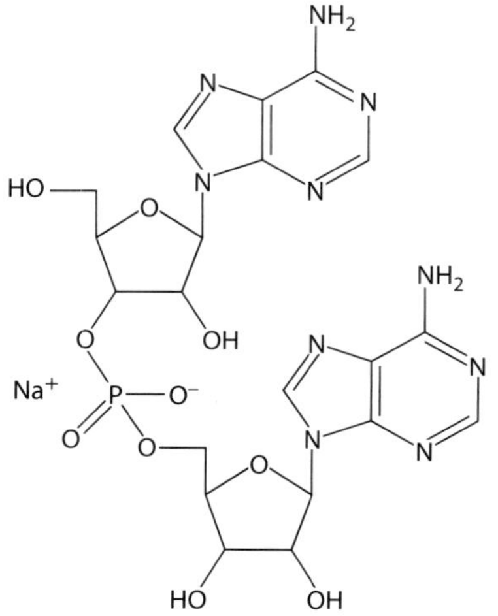

FIGURE 3.4 ApA dinucleotide.

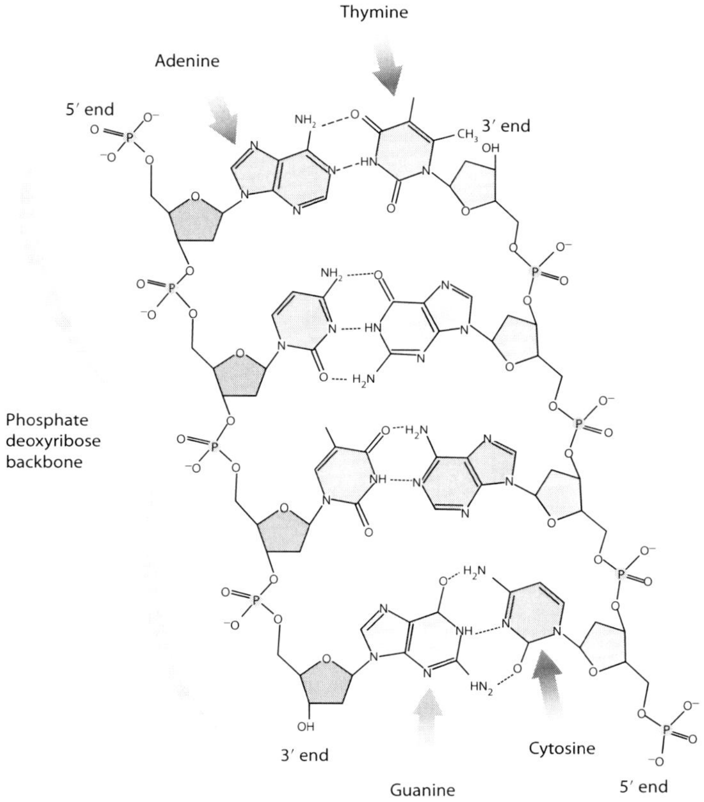

FIGURE 3.5 **(See color insert.)** Nucleic-acid base pairs. (Adapted and modified from https://en.wikipedia.org/wiki/File:DNA_chemical_structure.svg.)

sequences of bases are complementary to those of the first chain, a double-helical structure of the nucleic acids is formed (Figure 3.6). We mentioned earlier that the base–sugar moiety or nucleosides of the polynucleotide chain are hydrophobic. Thus, in an aqueous medium, they repel water and two polynucleotides orient themselves toward each other. Hydrogen bond formation between complementary bases gives a final stability to the helix; this is achieved by molecular recognition. A hydrophobic interaction between the polynucleotide and water is the main driving force for double-helix formation.

DNA can have three different structures: right-handed A and B and left-handed Z-DNA. In an aqueous medium consisting of 90% or more water, DNA exist predominantly as B-DNA. However, with a decrease in

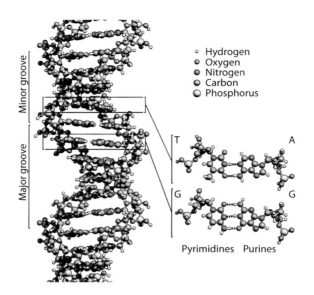

FIGURE 3.6 **(See color insert.)** Right-handed B DNA. (https://upload.wikime-dia.org/wikipedia/commons/4/4c/DNA_Structure%2BKey%2BLabelled.pn_ NoBB.png.)

water content in the medium and an increase in G–C base pairs, they form an A-DNA-like structure. The A-DNA is also right-handed, but base pairs are tilted more toward the helix axis. The diameter of the helix is also greater in comparison with that of the B-DNA. Double-stranded RNA, such as ribosomal RNA, adopt an A-DNA-like structure.

It has been recently shown that microorganisms thrive in extreme environmental conditions, where the DNA of the microorganism exists in A-DNA like structure. These organisms are known as extremophiles. The A-DNA here is wrapped around by a single alpha-helix protein. Cryoelectron microscopy at 4 A° resolution was used to determine the structure. Due to this structural element the organism can survive at pH3 and temperature as high as 80°C.

3.3.2 DNA Origami (Dynamic Molecular Recognition)

The hierarchical assembly of DNA chains utilizing base-pairing schemes and blunt-end stacking interactions gives rise to visually impressive and useful structures, as shown in Figure 3.7. Biomolecular complexes such as microtubules or actin filaments have large number of subunits and they are formed by dynamic recognition. They are weak and can be dissociated

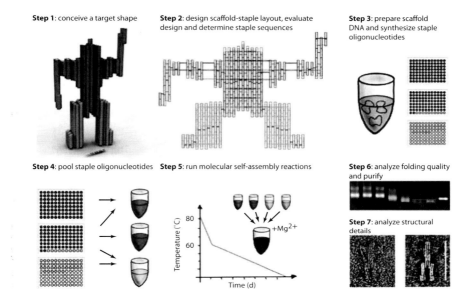

Step 1: conceive a target shape

Step 2: design scaffold-staple layout, evaluate design and determine staple sequences

Step 3: prepare scaffold DNA and synthesize staple oligonucleotides

Step 4: pool staple oligonucleotides

Step 5: run molecular self-assembly reactions

Step 6: analyze folding quality and purify

Step 7: analyze structural details

FIGURE 3.7 **(See color insert.)** DNA origami. (Carlos, C.E. et al., *Nature Methods*, 8(3), 2011.)

by a small amount of energy. DNA nanostructures are similar, and stacking interactions at the end of the chain are weak. However, the large surface area at the interface controls the stability of the assembled molecule. The desired configuration can be achieved by one master strand of DNA and numerous short staple strands, which can bind the master strand with Watson–Crick specificity. In this manner, 2D or 3D structures can be built by folding. The validity of these structures can be confirmed by both microscopy and gel analysis, as shown in Figure 3.7. These structures possess the potential to be applied as biosensors.

3.3.3 What Are the Major Recognition Points in DNA/RNA Helices?

Nucleic acid helices comprise three basic units (phosphates, sugars, and bases), all of which could serve as recognition points for different purposes. Phosphates can noncovalently bind to positively charged molecules, such as metal ions and basic amino acids; sugars can participate in hydrophobic interactions; and bases can interact by utilizing their aromatic properties. At times, macromolecular recognition between nucleic acids and proteins or between nucleic acids and carbohydrates involves all types of recognition groups.

It is interesting to know that both sugars and bases can exhibit minor conformational variations with drastic consequences with regard to molecular recognition. Sugars can have endo- and exo-conformations due to different puckering functions, whereas bases, in spite of being planar and aromatic, can undergo several conformational changes, as shown in Figure 3.8.

Sugars: Carbon atoms are tetrahedral in nature and when they cyclize, they can give rise to a different sugar geometry, as shown in Figure 3.9.

If one considers that C-1'-oxygen-C-4' is in one plane, then C-2' can move either up or down the plane, pushing the C-3' concomitantly in the opposite direction. Such a variation along with the rotation of the glycosidic bonds is a necessary determinant of the left-handed DNA.

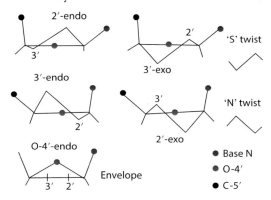

FIGURE 3.8   **(See color insert.)** Puckered structures of ribose sugars. (Adapted and modified from fbio.uh.cu.)

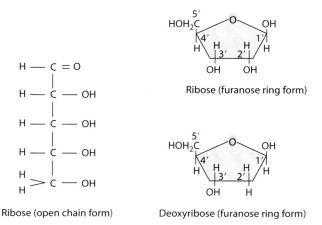

FIGURE 3.9   Different geometry of sugars.

Bases: H-bonded base pairs are planar (Figure 3.10). However, in the background of the sugar–phosphate backbone, they can adopt certain restricted conformational motifs that are important for protein recognition. Some of these motifs are shown in Figure 3.11. One can see in Figure 3.11 that the basic planarity rules of the aromatic base pairs are violated here, as H-bonds between base pairs are either twisted or staggered. Such a strained orientation can also be noticed between two successive base pairs.

Glycosidic bonds: The single bond between C-1' of the sugar and N-9 of purine or N-3 of pyrimidine is known as a glycosidic bond. The very nature

FIGURE 3.10   G-C base pairs in planar representation. (Adapted and modified from https://en.wikipedia.org/wiki/DNA#/media/File:Base_pair_GC.svg.)

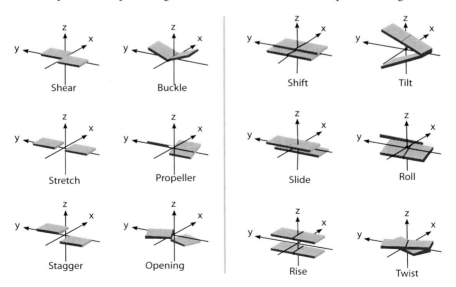

FIGURE 3.11   **(See color insert.)** Base pairs in altered conformation. (Adapted and modified from http://x3dna.org/.)

of the single bond enables it to have free rotation. However, due to the participation of neighboring groups, the energy levels at each rotation state are different, giving rise to *cis* and *trans* geometry, as shown in Figure 3.12.

As mentioned earlier, the geometry of the glycosidic bond and a variation in sugar puckers result in a global conformational change in DNA.

Grooves: The right-handed B-DNA has two grooves, major and minor (Figure 3.6), and the left-handed Z-DNA has one uniform groove (Figure 3.13). Many proteins noncovalently recognize these grooves and play a major role in gene expression. Since two major grooves lie on either side of the double helix with a defined number of bases between them,

FIGURE 3.12   Cytosine in DNA chain in trans orientation. (Adapted and modified from http://chemwiki.ucdavis.edu/.)

FIGURE 3.13   Left-handed Z-DNA. (http://molvisual.chem.ucsb.edu/.)

they play important roles in the recognition process. At times, a protein molecule wraps itself around the helix, facilitating the recognition of two adjacent major grooves (Figure 3.6). One can immediately understand that a Z-DNA binding protein cannot bind a B-DNA protein with specificity due to the change in groove dimension.

Stacking: The basic principles of molecular recognition are well exemplified in the stacking interaction involving DNA base pairs. An amino acid containing aromatic side chains can stack well by an insertion of the aromatic ring inside the *pi*-electron cloud of two successive base pairs. This is called *intercalation*. Similarly, a small inorganic complex such as cisplatin (Box 3.2) intercalates between two base pairs and interferes with DNA replication. The distance between two *cis* positions in square planar Pt(II) complexes is 3.4 Å, which is the distance between two successive base pairs in right-handed DNA, thus favoring intercalation. This mode of recognition has been successful in utilizing cisplatin as an anticancer drug.

Proteins: In biological species, the most important molecular recognition takes place between the DNA and the protein molecules. Such sequence-specific interactions are spatiotemporal in nature; guide the development and speciation of organisms; and are manifested throughout their lifetime. In Chapter 1, we discussed the constituents of protein molecules and the structure of a variety of amino acids. Amino acids are of various types: hydrophilic, hydrophobic, acidic, basic, and aromatic. The structure and the spatial disposition of the side group interrogate the DNA backbone, which, ultimately, results in the recognition process.

In the previous section, the various structural parameters that are important for recognition processes are depicted. Here, the major 3D structures

BOX 3.2 STRUCTURE OF CISPLATIN

Pt(II) has d^8 electronic configuration and forms square planar dsp^2 hybridized complexes. As shown in the accompanying figure, dichlorodiaminePt(II) can have two geometries, cis and trans. In the cis form, the distance between the adjacent positions is close to 3.4Å, which is

Source: http://www.ofichem .com/en/cytostatics-cisplatin/

the distance between two successive base pairs in B-DNA. Thus, the cis form of the Pt(II) complex or cisplatin can easily "fit in" the DNA or intercalate with the simultaneous removal of two chloride groups. In trans configuration, this is impossible, as the leaving chloride groups are far apart.

of proteins identified so far in biological samples are presented. These structures are alpha-helix, beta structures, random coil, helix-turn-helix, Leucine zippers, and Zn-fingers. They are very common in DNA-binding proteins.

Alpha-helix: Can a polypeptide chain fold to a regular structure? The peptide bond has a double-bond nature due to the H-shift from amide to carbonyl; thus, the conformation of the side groups is restricted around the peptide bond. Interestingly, only a few generalized conformations are possible due to steric constraints. The alpha-helix is one such conformation; it possesses a rod-like structure that is tightly coiled, and its side chains protrude in a helical array (Figure 3.14).

In this structure, the CO group of each amino acid forms a hydrogen bond with the NH group of another amino acid that is placed four residues ahead. Thus, all amino and carbonyl groups of this structure are hydrogen bonded except near the end. The polypeptide sequence in the alpha-helix is coiled around the longitudinal axis in a right-handed fashion while exhibiting a unique periodicity, and each amino acid is related to its nearest neighbor in a similar manner. The angle of rotation around the

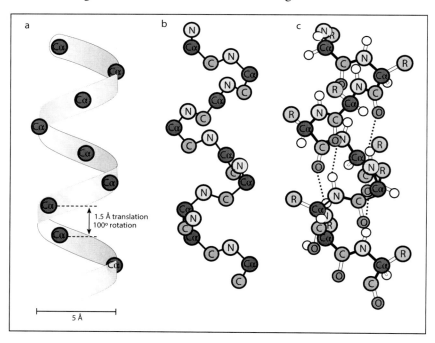

FIGURE 3.14 **(See color insert.)** Alpha-helix. (Adapted and modified from http://cmgm.stanford.edu/.)

peptide bonds between two successive amino acids is 100°, and the translation along the helix axis is 1.5 Å. Thus, to attain a full turn of 360°, there should be a noninteger number of amino acids. This unique concept was initially visualized by Linus Pauling (Box 3.3). Since the peptide unit is planar (Figure 3.15) due to the double-bond nature of the NH-C=O bond,

BOX 3.3 LINUS PAULING

Born in the state of Oregon in the United States in 1901, Linus Pauling was not only a great chemist but also a fine biochemist, a peace activist, an educator, and a writer. His contribution to chemistry during the 20th century remains unparallelled. His classic "The Nature of Chemical Bond" is widely read by students throughout the world. Pauling taught at Caltech and was one of the founders of Quantum Chemistry and Molecular Biology. Pauling was a two-time Nobel Prize winner: Chemistry (1954) and Peace (1962). He proposed the secondary structure of proteins and was also the first person to describe the molecular basis of the disease sickle-cell anemia. During the later stages of his life, Pauling advocated for the use of vitamin C as the cure for common cold and cancer.

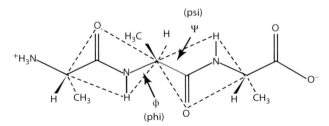

FIGURE 3.15 A dipeptide unit. (Adapted and modified from http://www.iop. vast.ac.vn/theor/conferences/smp/1st/kaminuma/UCSFComputerGraphicsLab/ AAA.html.)

it is possible to achieve only restricted conformations. Two angles ϕ and Ψ, describing rotations around the C_α-NH and C_α-CO bonds, respectively, have certain sets of values that are allowed. Other values will be disallowed due to a steric clash between side chains of the amino acids. These angles are known as *dihedral angles* (Box 3.4).

BOX 3.4 G.N. RAMACHANDRAN AND RAMACHANDRAN ANGLES

A trained physicist, G.N. Ramachandran (or GNR) was born in Ernakulam, India, in 1922. His contribution to conformational analysis of biopolymers, particularly of proteins, is ranked as an original contribution to structural biology. Regularly folded structures of biological macromolecules were expected since the description of the alpha-helix by Pauling (1951) and the elucidation of the double-helical structure of DNA by Watson and Crick (1953). From 1954 to 1955, Ramachandran  published the triple-helical structure of collagen through which he firmly established the ordered nature of proteins. However, Ramachandran realized that there was little or no information available on the possible conformation of polypeptide backbones. To address this problem, Ramachandran along with his colleagues constructed a map called the ϕ-Ψ Map (Ramachandran Map) or the Ramachandran Diagram.

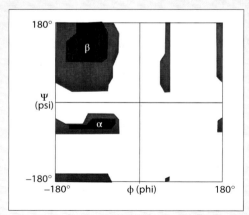

Source: http://www.iop.vast.ac.vn/theor/conferences/smp/1st/kaminuma/ UCSFCom-puterGraphicsLab/AAA.html.

Beta sheets: These are fully extended structures (Figure 3.16) that consist of polypeptide chains lying adjacent to each other in a parallel or an antiparallel orientation. Most of the proteins have both alpha-helical and beta-sheet arrangements in their peptide backbone and together, they participate in molecular recognition.

Helix-turn-helix: There are additional structural elements in proteins, such as helix-turn-helix, via which two alpha helices are joined by a specific sequence of amino acids. These helices can recognize two adjacent grooves of DNA in a sequence-specific manner (Figure 3.17). Transcription factors and repressor proteins utilize these modes of interaction to recognize DNA enhancers or operator elements. One such helix is known as the *docking helix*, whereas the other helix is called the *recognition helix*. The phages lambda CI repressor and Cro repressor possess these structures and exhibit binding activities to different operator sites on DNA.

Zn-fingers: Zn-finger–containing motifs in proteins were first discovered in the transcription factor regulating the transcription of the 5S RNA gene. The amino-acid sequences of such proteins have cysteine and histidine residues at regular intervals, and two of each coordinates with a central Zn(II)

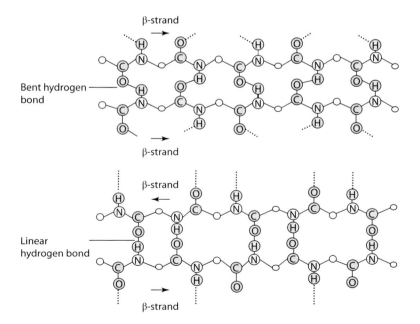

FIGURE 3.16   **(See color insert.)** Parallel strands with bent H-bonds and antiparallel strands with linear hydrogen bonds.

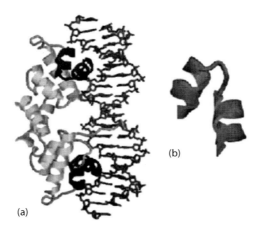

(b)

(a)

FIGURE 3.17 (a) Recognition of DNA by a helix-turn-helix protein. Helices are marked black and recognize two major grooves. (b) Helix-turn-helix motif. (Jones, S., Barker, J.A., Nobeli, I., and Thornton, J.M., *Nucleic Acids Res.*, *31*(11), 2811–2823, 2003.)

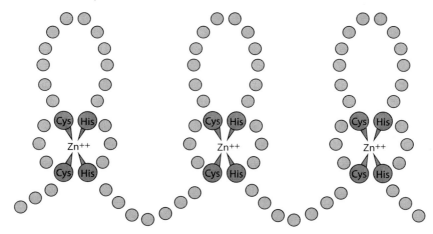

FIGURE 3.18 Zn-finger motifs. (Adapted and modified from http://genes. atspace.org/.)

(Figure 3.18) result in finger-like motifs. The basic amino-acid residues at the tip of the finger act on the G–C rich sequence of DNA, thus giving rise to specificity. However, several variations are now available for finger proteins.

Leucine zippers: A periodic repetition of leucine amino acids at every seventh position of an alpha-helix generates this motif. It helps in dimerization of the alpha helices to a chopstick-like structure (Figure 3.19) and

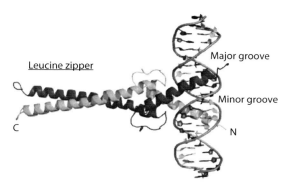

FIGURE 3.19 bZIP dimer is bound to DNA fragment: each α-helix represents a monomer. (https://en.wikipedia.org/wiki/Leucine_zipper#/media/File:Bzip_wikimedia_modified.tif.)

enables protein–protein recognition. If the helices are modified with basic amino acids at their tip, they can perform the DNA recognition function. Thus, this interdigitated motif has a great influence on molecular recognition for protein–protein oligomerization as well as on DNA recognition by a transcription factor.

So far, we have attempted to see major recognition functionalities in both DNA and proteins and their structural motifs. However, in order to understand the process that leads to recognition, one must understand the complementarities of shapes in biological macromolecules.

3.3.4 Proteins That Bind Nucleic Acids

Basic proteins constitute the most important group of proteins that recognize nucleic acids. These proteins primarily bind negatively charged phosphates of the backbone of nucleic acids in a sequence-independent manner. They promote the stability of the structure of nucleic acids, and this can be measured by an increase in melting temperature. In addition, the compaction of DNA by histones is an example of a typical histone–DNA interaction based on two oppositely charged species. The nucleosomal assembly process is one of the later steps in a histone–DNA interaction. Histones are lysine- and arginine-rich basic proteins and they have no sequence preference for DNA, with a few exceptions. In bacteria, there are many such proteins for example, HU, integration host factor (IHF) and DNA binding proteins from starved cells (DPS), to name a few. The DPS nonspecifically binds DNA and also exhibits a cooperative interaction. Such cooperativity augments the strength of binding. The x-ray structure analysis of DPS shows that the protein is multimeric in nature and, thus, resembles histones. When the interaction between a DPS and a nonspecific DNA was compared with the specific interaction between a promoter sequence and RNA polymerase (Figure 3.20), it was observed that the former was cooperative whereas the latter followed strict stoichiometry. However, the binding affinity was not significantly different.

Interestingly, the DPS from some bacterial species exists in two multimeric forms. Among these, one is the trimer form that does not recognize DNA. Its function is to titrate out free radicals generated by iron or by the Fenton reaction. Trimeric DPS from mycobacteria form a stable complex with Fe(II), as shown in Figure 3.21. On the other hand, a dodecamer, the other multimer, tightly wraps around DNA and protects it from free radicals and other harmful effectors inside the cell (Figure 3.21). This observation leads one to think whether the necessity of protecting the genomic material influences the structural organization of the protein.

In fact, DNA–DPS recognition gives rise to a different degree of organization of DNA as a function of growth of the organism. During the growth, the compaction of DNA within a small volume of the cell is necessary and it can best be achieved by an interaction with an oppositely charged protein. The compaction of DNA by an oppositely charged protein

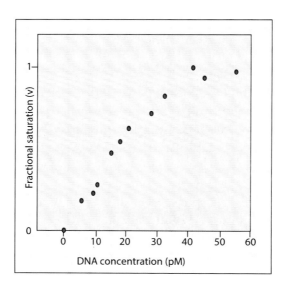

FIGURE 3.20 Plot of fractional saturation of RNA polymerase with promoter DNA.

is important; there are several ways in which such compaction can take place, as shown in Figure 3.22. The central theme in all these cases is molecular recognition.

Fe^{2+}-DPS

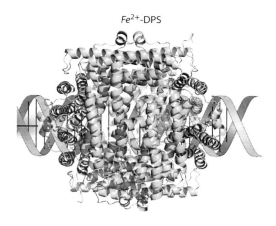

FIGURE 3.21 **(See color insert.)** Iron-bound DPS. (Arnold, A.R., and Barton, J.K., *J. Am. Chem. Soc., 135*(42), 15726–15729, 2013.)

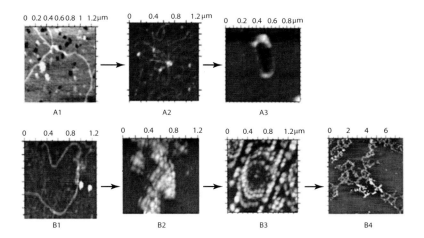

FIGURE 3.22 DPS and the nucleoid organization in Mycobacterium smegmatis. A: Nucleoid organization by DPS 1 from free DNA to toroid. (1) Free DNA, (2) DPS1-DNA in log phase (24 h), and (3) DPS 1-toroid in late stationary phase (144 h). B: Nucleoid organization by DPS 2 from free DNA to coral reef. (1) Free DNA, (2) DPS 2-DNA complex in log phase (24 h), (3) DPS 2-toroid in late stationary phase (72 h), and (4) coral reef structure in very late stationary phase (144 h).

HU and IHF, both of which are DNA-bending proteins, recognize DNA as possessing unique structural parameters. IHF binds DNA through a minor groove and destroys a stacking interaction within DNA base pairs by inserting a proline residue. IHF is known as a master bender and it bends DNA by 160°, thus reversing the direction of the helix.

Another DNA-bending protein is cyclic adenosine monophosphate (c-AMP) receptor protein (CRP), which acts as a transcription activator. This protein has the unique ability to recognize the ligand c-AMP at one end, whereas the other end binds DNA. Two CRP molecules act together, and the ligand binding is cooperative in nature. Once the DNA is bound at the major groove, it bends the molecule by almost 90° and opens up the promoter region, thus facilitating RNA polymerase recognition. This again points to the fact that the necessity for compaction and the packing of DNA are guiding the recognition between two partners, DNA and protein.

The questions that would be asked are obvious. What are the functional implications of sequence-specific recognition between two macromolecules? What are the steps that lead to the sequence-independent DNA–protein recognition? These questions will be discussed in the next section on how different mechanisms operate for sequence-dependent recognition between DNA and proteins. However, sequence-independent recognition is a structural requirement.

Protein folding is mainly dictated by its primary amino-acid sequence, and the native structure always attains minimum free energy conformation. Subsequently, a DNA-binding protein will recognize the phosphate backbone of nucleic acids. To accommodate the interacting protein, the DNA backbone will be modified by bending, thus producing kinks or opening up the helix. This binding can be stoichiometric or allosteric and could involve many molecules, as mentioned earlier for DPS. However, the nonspecific interaction between DNA and protein is a structural requirement that provides stability to the system. At times, it is more important than the control of gene expression, which is the outcome of a sequence-specific interaction between DNA and a protein.

3.3.5 Sequence-Specific Recognition between DNA and Proteins

Here, we will consider two kinds of DNA sequences: an operator and a promoter. There are other sequence elements such as enhancers in higher organisms; however, their mode of interaction is similar to that of promoters, and they possess additional parameters such as DNA bending. We will discuss a few such examples of the same.

"A gene is defined as the sequence of bases that is represented in a diffusible product." By the same token, an *operator* or a *promoter* can be defined as the sequence of bases that is not represented in a diffusible product. However, both operators and promoters regulate gene expression by binding to different proteins with sequence specificity and, most often, with one binding site. This is one of the major differences from the group of proteins discussed earlier.

What are the other differences between operator and promoter sequences? An operator is represented by a twofold symmetry element or it is said to be palindromic in nature, whereas a promoter is not. As a consequence, an operator is recognized by a repressor protein that is multimeric, and it has an overall twofold symmetry in a 3D structure. On the other hand, the cellular transcription machinery, the RNA polymerase, is also multimeric but has no structural symmetry. It recognizes the promoter site, specifically resulting in unidirectional movement of RNA polymerase and transcription. Both operator and promoter together regulate gene expression by recruiting repressor or RNA polymerase, respectively. At times, the operator and the promoter element in DNA overlap and due to steric constraints, the binding of the repressor or RNA polymerase is mutually exclusive, resulting in important biological consequences.

When a gene needs to be transcribed, RNA polymerase recognizes the promoter element and the mechanism is known as a *promoter search*. Usually, this is the rate-determining step in RNA synthesis. The DNA-bound RNA polymerase is known as the closed complex (RPc), as the DNA base-pair opening has not yet taken place (Figure 3.23). Subsequently, DNA melting, a prerequisite of transcription, occurs and the same complex will be called an *open complex* (RPo), which is characterized by a certain shift of RNA polymerase over DNA and the generation of single-stranded DNA. The open complex is necessary for the initiation of transcription.

However, at times, the transcription process needs to be regulated; further, depending on the environmental cue, the gene is turned either on or off. The requirement here is that the enzyme RNA polymerase should not occupy the promoter site. This can be achieved by blocking the access of the enzyme to DNA, which is a kind of nonrecognition. The most common method that could be utilized to activate this process is having a

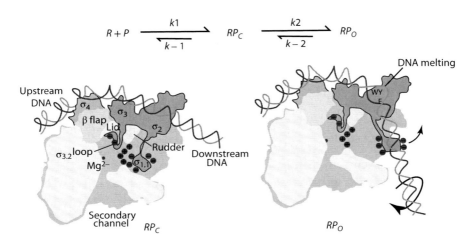

$$R + P \xrightleftharpoons[k-1]{k1} RP_C \xrightleftharpoons[k-2]{k2} RP_O$$

FIGURE 3.23 **(See color insert.)** Binding of *E. coli* RNA polymerase with T7A1 promoter. (Murakami, K.S., and Darst, S.A., *Curr Opin Struct Biol* 13(1), 31–9, 2003.)

symmetric operator site adjacent to the promoter. When gene transcription is either not required or redundant due to environmental signals, a protein repressor will bind very tightly to the operator site, thus blocking further interactions between the promoter and RNA polymerase. This is a simple manifestation of steric crowding. However, when the signal is removed or when gene synthesis becomes a necessity, the repressor will dissociate from the operator and gene transcription begins.

The recognition between an operator and repressor has also been worked out thoroughly by employing various structural means. It was observed that the helix-turn-helix motif is one of the common structural features of the repressor molecule. As shown in Figure 3.16, a helix-turn-helix protein wraps around the DNA tightly so that two alpha helices can be inserted into two successive major grooves in the DNA.

During the recognition process of DNA by protein molecules, aromatic amino-acid residues play a major role. A tryptophan or tyrosine residue can be inserted between two aromatic base pairs in DNA, in much the same manner as done earlier in the interaction between IHF and DNA. Such *pi*-stacking gives stability to the interacting pairs and is also a common mechanism used for nucleic acid–protein recognition.

Zn-fingers: In Section 3.3.3, the DNA-binding motif of Zn-fingers was discussed. Zn(II) is a very useful metal ion in biology. Its usefulness lies in the fact that unlike transitional metal ions, Zn(II) is not very reactive due to its electronic configuration. The most common oxidation state of Zn, Zn(II) has filled d-orbital ($3d^{10}$) and is also optically inert. However, structurally, it takes part in many enzymatic functions. The estimation of Zn in a typical protein is often very difficult due to the lack of any spectroscopic signals. However, one can use a radioactive isotope, ^{65}Zn, or a fluorescent ligand specifically meant for Zn estimation. The best method employed so far for Zn determination, however, is atomic absorption spectrometry; this method can estimate Zn at a 2 ppb level. At times, the stoichiometric level of Zn to protein increases to more than one or two Zn atoms per protein molecule. This helps a great extent in understanding the biology behind the DNA recognition pattern of the protein.

Almost all the Zn-finger proteins have tetra-coordinated Zn, which is surrounded by the amino acids cysteine and histidine. These bind the central Zn in different proportions from all four cysteines to all four histidines. The first Zn-finger protein TFIIIA was discovered in the transcription of the 5S RNA gene of *Xenopus laevis*; it has nine Zn-finger domains per protein molecule. All the finger domains jointly bind the promoter-like sequence of 5S DNA. However, the promoter is placed within the DNA itself and is transcribed in a different manner from prokaryotic promoters. This promoter DNA sequence is known as an *intragenic control region*.

3.3.6   RNA–Protein Recognition

Ribosomes: Ribosomes are a large ensemble consisting of RNA and proteins. They perform protein synthesis, which is one of the most important functions within a cell, from an intermediate RNA molecule. They are also known as a *protein synthesizing factory*. Ribosomes from both prokaryotic and eukaryotic sources are made of several RNA and a large number of basic proteins. They link amino acids together as per the instructions from the base sequence available in messenger RNA. At this stage, the base-pairing scheme of recognition between nucleic acids is translated to polypeptide synthesis; this is made possible through the formation of a peptide bond. The small subunit of ribosome recognizes the amino-acids and they are linked together to form the peptide bond in the larger subunit (Figure 3.24). The entire process of peptide bond formation involves a small adapter molecule, tRNA, which has a nucleic-acid recognition motif at one end and an amino-acid recognition motif at the other end. In biology, it is rare to witness such a kind of dual recognition behavior.

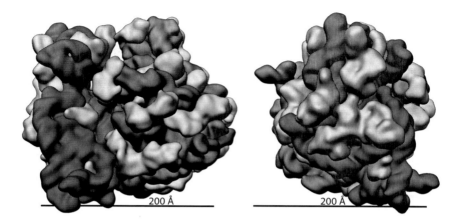

FIGURE 3.24   **(See color insert.)** Association of 50S (RED) and 30S (BLUE) subunits to form 70S bacterial ribosome. (https://upload.wikimedia.org/wikipedia/commons/4/42/Ribosome_shape.png.)

There are more than one RNA molecule in ribosome and more than fifty proteins which are bound together in noncovalent interaction, performing the finest enzymatic reaction like peptide synthesis. In addition, mechanical movements could be observed on ribosomes that result in the translocation of the growing peptide chains. The structure of the ribosome has been solved by x-ray crystallography and is known now at the atomic level (Box 3.5). Unlike messenger RNA (mRNA), the RNA molecule in the ribosome is double stranded and is known to have an A-DNA–like structure. Crystallographic analysis has also revealed that the proteins in the ribosome act as scaffolds; no proteins are located near the site where peptide bond formation takes place. Instead, RNA molecules act as enzymes that help in the catalytic process of peptide synthesis. Thus, the ribosome is also known as a *ribozyme*.

Although the ribosome is the best example of the RNA–protein complex, there are also other biologically important reactions that are controlled by specific RNA–protein recognition. Proteins interact with RNA by following a similar principle, such as DNA–protein recognition. These interactions are noncovalent in nature, electrostatic, and hydrophobic; stacking interactions between aromatic bases of RNA and amino-acid side chains dominate the recognition process. There are also cases of Zn-fingers in RNA-binding proteins that specifically recognize double-stranded RNA molecules. Double-stranded RNA are similar to A-DNA,

BOX 3.5 RIBOSOME STRUCTURE AND VENKI RAMAKRISHNAN

The general molecular structure of ribosome was attempted from 1970 onward, and initially, various methods such as Neutron diffraction and immuno-electron microscopy were employed to get an approximate idea of the structural organization of this ribonucleoprotein particle. A bacterial ribosome has three different RNA molecules and more than fifty proteins arranged in a precise order in two subunits: 30S and 50S. "S" denotes the Svedberg unit, which measures the rate of sedimentation and not size. The detailed functional domains in the ribosome, the growing polypeptide chain, and the peptide exit channels are shown in the figure given below. The ribosome assembly follows a particular scheme and was worked out in detail some time ago. However, crystal structure determination in high resolution was achieved only at the beginning of this century. The magnificent structure of the ribosome and its antibiotic binding sites were awarded the Nobel Prize in 2009. One of the recipients of the Nobel Prize, Venki Ramakrishnan, is an Indian physicist who was initially trained in India and then in the United States. Recently, Ramakrishnan demonstrated that cryoelectron microscopy can also be used to solve the high-resolution structure of the ribosome. Ramakrishnan is currently the president of the Royal Society, the United Kingdom.

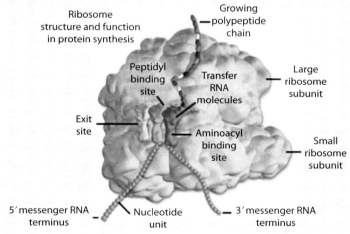

Source: http://neobiolab.com/research/wp-content/uploads/2015/10/Ribosomes-Protein-Factories-their-Role-Initiation.jpg.

and, thus, these proteins exhibit a specific pattern in recognizing A-like structures. Transcription factor TFIIIA, a Zn-finger protein, recognizes both DNA and RNA; the interaction takes place through an A-like motif.

Although RNA-binding proteins are mostly present in the cytoplasm, their presence in the nucleus cannot be ruled out. Before being transported to the cytoplasm, many pre-mRNAs are bound to proteins in the nucleus.

3.3.7 Structural Basis for Protein–RNA Recognition

RNA performs diverse functions within cells; it is also a part of many different, larger assemblies such as ribosomes and spliceosomes. Apart from its all-important biological functions, RNA is also known to exhibit catalytic activities. Small RNA molecules such as miRNA have a profound impact on gene regulation. However, structural information on RNA–protein complexes is limited when compared with that on DNA–protein complexes.

One of the interesting aspects of RNA–protein recognition is the diverse nature of the interacting surface between RNA and proteins. Van der Waal interaction is more predominant in this case than are H-bonds. G and U bases are more preferred surfaces in the RNA molecule, although they have two hydrogen bonds between them (Box 3.6).

BOX 3.6 NON-WATSON–CRICK BASE PAIRS

The canonical Watson–Crick base pair of DNA is a classical static representation of molecular recognition. However, it is increasingly being felt that it does not depict the actual dynamics. Transient sampling events of several different base-pairing schemes are simultaneously going on, such as G–T and G–U base pairs with two hydrogen bonds between them. At any given time, the concentration of such non-Watson–Crick pairs is sufficiently high to influence biological functions such as replication and translation. Recently, NMR experiments were carried out to identify such wrong pairs.

Wobble base pair   Enol tautomer of T

Source: Adopted and modified from http://dnamismatch.com/Test/wp-content/uploads/2013/07/The-G%C2%B7T-mismatch.png.

3.3.8 tRNA–Aminoacyl tRNA Synthetase Interaction

Earlier, we dealt with small adaptor molecules, that is, tRNA, which has both nucleic-acid recognition motifs and an amino-acid binding domain. These molecules are extremely important, as they decide the sequence of polypeptide synthesis or the integrity of the resulting protein. The triplet code of bases in the messenger RNA is recognized by another complementary triplet in tRNA (anticodon), and, thus, these tRNAs are different from each other. If 20 triplets are lined up sequentially, there will be 20 different tRNAs; each tRNA is specific for one triplet code. The codon–anticodon recognition is strictly noncovalent in nature; it follows the Watson–Crick base-pairing scheme, and the recognition sequence is antiparallel.

On the other end of the cloverleaf model of tRNA (Figure 3.25) is a triplet base CCA that binds acylated derivatives of amino acids. This is known as *aminoacylation*; it is a covalent modification and is enzymatic in nature. The nature of amino acid added at the CCA end is guided by the anticodon sequence of the tRNA. A true representation of "action at a distance" is operational here. Thus, there are 20 different aminoacyl tRNA synthetases; each one of them is specific for an amino acid and the triplet code. Importantly, aminoacylation is carried out with the aid of adenosine

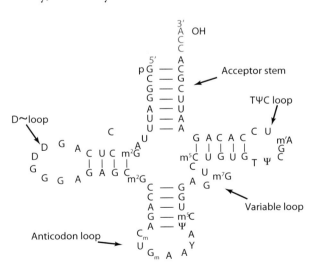

FIGURE 3.25 Secondary structure of tRNA. (Adapted and modified from https://upload.wikimedia.org/wikipedia/commons/thumb/5/59/TRNA-Phe_yeast_en.svg/220px-TRNA-Phe_yeast_en.svg.png.)

triphosphate (ATP), which donates the required energy. This adaptor molecule not only recognizes two different partners but also recognizes them as exhibiting two different kinds of bonding—one is hydrogen bonding and the other is covalent bonding.

It is important to notice that the mere presence of an extra -OH group at the sugar end results in innumerable, different biological functions of RNA. RNA is involved in evolution, and it is constantly breaking many dogmas related to biology. It is interesting to note that whenever RNA participates in a biological process, a protein is involved. This makes the studies on RNA–protein interactions extremely significant.

3.3.9 Recognition by miRNA (Micro RNA)

miRNAs are RNAs that are a little more than 20 nucleotides long; they are widely used to regulate gene expression. They are found in plants, animals, and viruses and they bind the complementary sequence of the mRNA molecule. On binding the complementary sequence of mRNA, they inhibit the translation process. Several such molecules have been discovered in the recent past (Box 3.7).

BOX 3.7   miRNA

The first miRNA was discovered in *Caenorhabditis elegans* while following the timing in larval development. The gene that was involved in the control of timing was the lin-4 gene. However, the effect was not mediated through a protein; rather, it was mediated by a 22-nucleotide-long small noncoding RNA that has partial sequence complementarities to some region of the lin-4 mRNA. These complementarities appear to control the expression of certain genes, primarily taking part in developmental transition and inhibiting translation. Within 10 years of this first observation, several such small RNAs were found in various other organisms and it was believed that "small temporal RNAs" are the key elements for the timing of development. Currently, the function of these small RNAs is well established and is termed miRNA. They are mainly involved in regulatory pathways during gene expression.

miRNA is synthesized by RNA polymerase II as a precursor RNA of a longer length. It is usually transcribed as independent units and is oriented antisense to the neighboring genes. Pre-miRNA is exported from the nucleus to the cytoplasm, where it is processed by an RNase III, called Dicer.

3.4 POLYMERIZATION OF NUCLEIC ACIDS

Both DNA and RNA are polymerized in a template-dependent fashion strictly guided by the fidelity of the catalyst, which is DNA and RNA polymerase. The overall mechanism of phosphodiester bond formation is the same for DNA and RNA synthesis although there are differences in each step of the polymerization process. Both RNA and DNA first recognize the nucleic acid templates in a sequence-specific manner. These sequences are known as the *origin of replication* and the *promoter sequence* in the case of DNA and RNA, respectively. Both these recognition processes require the generation of a single-strand region over the nucleic acids, either at the recognition point or a little away from it, so that new bases can be added over the single strand. This addition is regulated by the base sequence of the single-strand region of the template. The enzymes select the nucleotide from the pool of substrates, as dictated by the template. Thus, the enzyme has three major functions: (1) recognition of the template in a sequence-dependent manner, (2) listening to the terms of the template, and (3) substrate recognition. The generation of the final dinucleotide is due to the formation of a covalent bond between two substrates that are guided by the SN2 (Substitution Nucleophilic, bimolecular) mechanism, as shown in Figure 3.26.

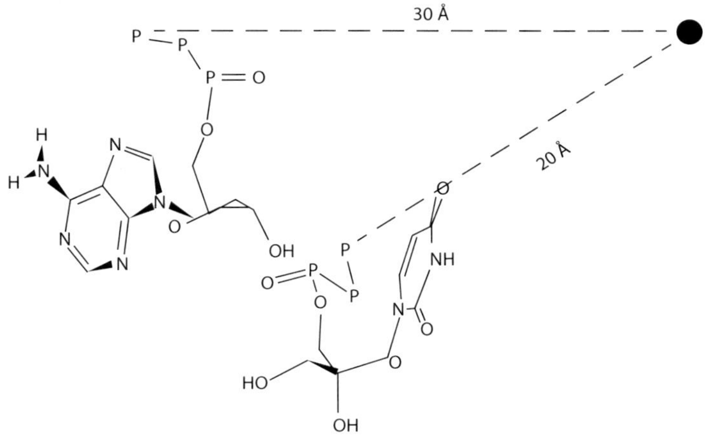

FIGURE 3.26 Nucleic-acid polymerization. The orientation of two nucleotides with respect to each other during the formation of the first phosphodiester bond at the RNA polymerase active site. The dark circle represents the antibiotic rifampicin-binding sites, and the distances are actually measured by fluorescence spectroscopy. The attacking group, 3'-OH and the electrophilic center alpha phosphorus lie in one plane.

The polymerization of both RNA and DNA in the biochemical sense is thermodynamically unfavorable; further, it is carried out in the presence of an enzyme catalyst at a physiological temperature. The polymerization triggers the process while exhibiting a high degree of specificity. Both DNA replication and transcription demand that there are at least two specific recognition points for the enzyme: one with the template (nucleic acid–protein recognition) and the other with the substrate (nucleotide–protein recognition).

During the formation of the first phophodiester bond, there are two substrates: the initiating nucleotide (i-nucleotide) and the second nucleotide (i+1-nucleotide). The 3'-OH of the first nucleotide is de-protonated by a metal atom, and the resulting nucleophilic center (O-) attacks the innermost phosphorus of the incoming NTP (i+1) nucleotide. The polarization of the orthophosphate generates a partial positive charge at the phosphorus. As mentioned earlier, the attacking group and the leaving group, PPi, are lying opposite to each other and are collinear. This is the primary requirement of the SN2 attack (Figure 3.27). During polymerization, the same process continues for the subsequent addition of nucleotides. It should be noted that the terminal phosphates for the first nucleotide at the 5'-end play no role in phosphodiester bond formation. For the subsequent bonds, the phosphates serve as only the electrophilic center and the leaving group.

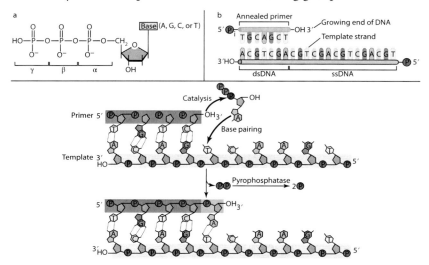

FIGURE 3.27 **(See color insert.)** Chemistry of DNA synthesis. (Pearson Education, Inc., http://www.slideshare.net/narayanprahlad/dna-replication-and-enzymes-involved-in-dna-replication.)

In the case of DNA replication, two daughter strands are generated from the parent DNA helix by double-strand opening; one of these strands is faithfully copied. In addition, several other proteins also participate in this process. The enzyme DNA polymerase has both template recognition and substrate recognition motifs. Either other variants of the same enzyme could correct the wrong incorporation or an inherent error-correction mechanism could be employed.

The beauty of the molecular recognition manifests itself best in each step of the central dogma. Apart from nucleic acid–protein recognition, which is sequence specific in nature, substrate nucleotide binding to the enzyme is specific and is directed by the sequence of the template. The selective choice of a nucleotide from a large pool within the cell is remarkable, as the consequence of a mistake can be detrimental for the survival of the organism. However, during the process, if a mistake occurs, it can be corrected. Such a correction involves the second level of the recognition process between the nucleic acid and the protein.

The process of DNA polymerization has been recently visualized in a time-dependent manner. One may refer to it as *DNA polymerase in action*. Different snapshots of the substrate–enzyme complex by x-ray crystallography for a relatively slow DNA polymerase as a function of time reveals a similar mechanism of phosphodiester bond formation as discussed earlier for RNA polymerase. Figure 3.28 shows such a process.

RNA synthesis takes place using DNA as a template with a different enzyme, DNA-directed RNA polymerase. This process is known as transcription. In this case, the enzyme is much bigger and presumably performs more complicated functions. It needs to get direction from the DNA template in an error-free manner, and this is diverted toward substrate recognition. Then, the enzyme synthesizes the phosphodiester bonds and the process continues until the termination of RNA synthesis takes place.

Various subunits of multimeric RNA polymerase have different functions during the transcription process. In a bacterial species, RNA polymerase comprises $\alpha_2\beta\beta'\omega\sigma$; the core enzyme $\alpha_2\beta\beta'\omega$ is capable of RNA synthesis but recognizes DNA nonspecifically. On the other hand, the free sigma subunit, σ, does not bind DNA itself; however, it imparts sequence-specific DNA recognition on binding to the core enzyme. This process is known as *promoter recognition*. The DNA-binding domain of free sigma at the N-terminal is occluded by the C-terminal through intra-domain protein–protein recognition, which is a well-known control mechanism in biology. On binding

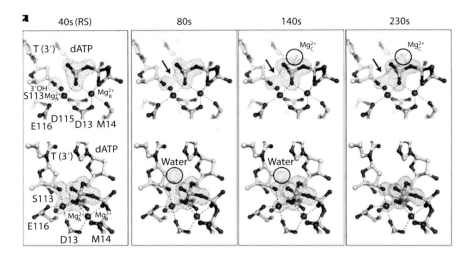

FIGURE 3.28 **(See color insert.)** Reaction time course of DNA polymerase. The reaction process is monitored by the electron density corresponding to the new chemical bond in the map of the reactant state (RS) compared with the refined GS (ground state). The emerging densities are pointed out or circled. RS is the reactant state structure where there is no sign of bond formation. Two unexpected spheres of electron density appear in the course of new bond formation. The first is within hydrogen bonding distance of the 3'-OH. Its electron density peaks at 80s and drastically declines after 140s when the PS (product state) becomes prominent. Water is hydrogen bonded with the 3'-OH, the O4' of dATP, and water in the 80s RS structure. The second emerging sphere of electron density is very close to the α-phosphate on the opposite side of the A- and B-site Mg2+ ions. It starts appearing at 140s and intensifies with the reaction time. Without an A-site divalent cation, the 3'-OH of the primer strand shifts away from the dATP and forms hydrogen bonds with the sidechains of S113 and D115. The active-site carboxylates D13 and E116 adopt different rotamer conformations. (Reprinted by permission from Macmillan Publishers Ltd. Nature, Nakamura, T. et al., Watching DNA polymerase η make a phosphodiester bond, 487, 196–201, copyright 2012.)

to sigma, core RNA polymerase releases the N-terminal domain. Various sigma proteins within the cell compete for the same core enzyme under different environmental conditions. At an elevated temperature, heat-shock sigma factor binds to the core and transports the enzyme specifically to the heat-shock promoter. This enables the heat-shock–responsive elements to be synthesized and facilitates the survival of the organism. Heat shock sigma containing RNA polymerase prefers heat shock promoters over other type of promoters in order to facilitate transcription of heat shock responsive genes.

The principles of molecular recognition, however, is same as that of RNA polymerase binding to any other promoters. The sigma cycling is strictly followed as per the requirement of the cell. Such will be the case when other sigma factors bind to core RNA polymerase as per the requirement of the cell.

Another kind of specific molecular recognition during transcription is evident through the activation mechanism. The basal level of transcription could be attained with promoter recognition, initiation, elongation, and termination. However, the activation and repression of transcription are the controls that are necessary for gene function. Such control of transcription occurs and is manifested by protein–protein recognition of activator/represssor with RNA polymerase. In Section 2.4, we discussed how the repressor binds the operator sequence and sterically hinders RNA polymerase binding to the promoter; the aim of this is to downregulate the transcription of those genes that are required to remain silent in the absence of the inducing sugar. Here, the operator and the promoter sequences are placed adjacent to each other. Similarly, during the activation of gene transcription, the activator-binding sequence could be adjacent to the promoter, however, at the upstream side without hindering the movement of RNA polymerase. On binding to cAMP, the CRP undergoes a conformational change and binds to DNA. As shown in Figure 3.29, this activator protein

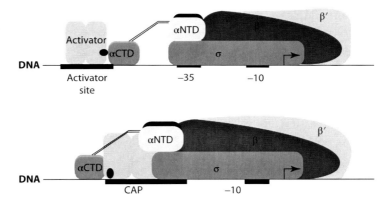

FIGURE 3.29 Binding of CAP protein to the promoter. The dimeric CAP protein is labeled "Activator." Binding to a class I promoter is shown in upper panel and binding to a class II promoter is shown in lower panel. The two classes of promoters are defined as the ones that require contact of the alpha subunit with the activator (class I) and both alpha and sigma subunits (class II). (Adapted and modified from http://www.biochemistry.ucla.edu/.)

then touches (or perhaps pushes) the C-terminal domain of the α-subunit of RNA polymerase. This results in the rapid movement of RNA polymerase and the activation of gene transcription. The mechanism is operational even when the activator is bound several base pairs away. However, the binding sequence needs to be upstream from the promoter at a multiple of 10 base pairs, so that both are in the same phase. Contact is established between the activator and the α-subunit by the looping of the intervening DNA sequence.

CRP (also known as CAP) binds cAMP while exhibiting major conformational changes, and this complex then binds and bends DNA (Figure 3.29). Subsequently, CRP recognizes RNA polymerase and, thus, it has three recognition schemes: (1) a nucleotide, (2) DNA, and (3) a protein. In all cases, recognition results in important biological and structural consequences.

3.5 MOLECULAR RECOGNITION AT THE HEART OF THE CENTRAL DOGMA OF MOLECULAR BIOLOGY

The central dogma of molecular biology that was enunciated by Francis Crick in 1958 forms the keystone of modern biology. This dogma deals with the detailed residue-by-residue transfer of sequential information from DNA to protein through an RNA intermediate. More than 50 years of work on each step of the central dogma, including DNA replication, transcription to intermediate RNA, and, finally, translation to protein, has taught us that specific molecular recognition due to noncovalent interactions between different players guides these processes. Currently, it is obvious that the machineries involved in transcribing and translating genetic messages have associated proteins with distinct functions. The reasons for such special benefits gifted to a particular metabolic path are quite obvious. Each macromolecular complex, such as nucleosome, RNA polymerase, ribosome involved in replication, transcription, and translation, respectively, is the mixture of proteins and nucleic acids. Their spatial disposition gives rise to a special shape that could be due to their special function.

In Section 1.4, we mentioned that structural complementarities play a very major role in macromolecular recognition. Therefore, folding of the protein in an accurate manner aids in molecular recognition. This step in protein folding is often guided by other proteins. The proteins that assist proper folding are known as *chaperones*. It has been recognized that the machineries participating in different steps of the central dogma have associated protein-folding activity. Well-known chaperones constitute the proteins that take part in the DNA replication complex. In the case of both lambda DNA and *Escherichia coli* DNA, chaperone proteins recognize a special sequence in DNA; this is known as the *origin of replication*. However, it is not clear whether any other function is deemed necessary with regard to these proteins to start the replication process. The protein-folding activity is not immediately discernible for the initiation of replication.

However, it is now clear that the assembly of RNA polymerase takes place after proper folding of the largest subunit of RNA polymerase (β') by its smallest counterpart, which is the omega subunit. The enzyme assembles to an active component in a stepwise manner; any wrong addition of a subunit will generate an inactive assembly. The entire assembly process is shown in Figure 3.30.

Various techniques are employed to follow the assembly of macromolecules; most of them are based on the recognition principle. Some of these methods will be discussed in Chapter 4 of this book. However, it needs to be appreciated that multisubunit enzyme assembly is a fascinating problem with regard to protein–protein molecular recognition.

Proteins are synthesized vectorially from their N-termini to C-termini on the ribosome, and this process is known as *translation*. They also undergo cotranslational folding as they extend from the peptidyl-transferase center through the ribosome exit tunnel, which is 100 Å in length. The nascent peptide is sheltered from external proteins and proteases and is offered a conducive environment for folding in this exit tunnel. Structures higher than a helix cannot be formed inside this tunnel, as it has a width of only 15 Å.

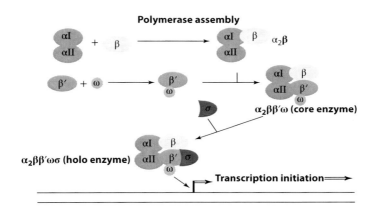

FIGURE 3.30   Assembly of bacterial RNA polymerase in a stepwise fashion.

In addition to providing a sheltered environment, evidence suggests that the ribosome itself can chaperone protein folding. In a wide variety of systems, the ribosome can effectively assist the folding of a protein according to its biological shape. Thus, the large RNA of the ribosome exhibits this intrinsic activity.

Similar to RNA polymerase, the stepwise assembly of the ribosome is very important for biological activity. However, the system is more complicated, as many proteins and RNA molecules serve as active players. Any wrong addition would both alter the recognition pattern and destroy the activity of the macromolecular complex.

3.6 POLYMERIZATION OF BIOLOGICAL MONOMERS

Directionality and shape constitute the most important part of polymerization. The process of polymerization is governed by the specific recognition of two or more macromolecules, which are most often proteins and nucleic acids. The process is thermodynamically unfavorable and can take place at room temperature and pressure, but only in the presence of enzymes.

Proteins or peptides are formed by a condensation reaction between two or more amino acids; at each step of the reaction, a water molecule is eliminated. The sequence of the addition of amino acids is the most crucial and is dictated by the genetic code embedded in DNA. In Section 3.3, this concept was adequately discussed. Two important aspects need special mention here. The sequence will guide the stability and the shape of the protein, which, in turn, will determine its function. On the other hand, chemical macromolecules are random and the sequence of monomers plays a very minor role. The fundamental reason perhaps is the lack of variability in the monomer units and the lack of directionality. If there are 20 different monomers such as amino acids that polymerize to generate a chemical macromolecule, the role of the sequence will have minimum effect if it has no order. As a result, the shape of a chemical macromolecule does not play any significant role in its function. However, there are a few interesting exceptions and these will be discussed later in Chapter 5.

Another aspect of importance in molecular recognition is the sequence of a series of phosphodiester bonds that are translated into a sequence of amino acids consisting of entirely different structural elements. This is possible because of tRNA, an adaptor molecule that can recognize both a triplet of nucleotides and an amino acid simultaneously. Thus, two ends of the same molecule recognize two different structures that are interlinked. However, this kind of recognition is extremely rare and it performs a major function of protein synthesis while exhibiting different shapes and functions.

There are specific enzymes for nucleic-acid synthesis. The mechanism used is template dependent; however, the DNA base sequence directs the synthesis of a complementary strand over the copying strand with the help of an enzyme DNA polymerase. Thus, the enzyme here has two recognition principles: (1) it binds to a specific base sequence that needs to be copied and (2) it picks up a complementary nucleotide from the nucleotide pool. The latter process is often aided by a divalent cation. These are the two basic functions. In addition, other equally important activities are governed by the molecular recognition process, such as DNA unwinding, DNA bending, and error correction. The entire process involves multiple protein complexes for specificity.

In the case of RNA synthesis, the DNA-dependent RNA polymerase enzyme performs a similar function to DNA polymerase. The enzyme RNA polymerase recognizes DNA template, unwinds it to single strand, copies one strand by taking the message in the form of base sequence, and adds a series of nucleotides in a specific manner until another protein comes to help in terminating the synthesis of RNA. Thus, the enzyme exhibits template recognition and substrate recognition activities. In addition, multiple small molecules and proteins bind the enzyme–DNA complex and regulate the rate of RNA synthesis. The mechanism related to phosphodiester bond formation is similar in both cases. Nucleic-acid polymerization thus follows the same principle of recognition, which consists of only template change (in the case of reverse transcriptase, it is the RNA template and deoxy-nucleotide substrates).

One may notice that the principle of polymerization has multiple approaches for all biological macromolecules. The covalent bond formation between two successive monomers is associated with an elimination reaction, and it enables the substrate to recognize a specific catalyst (enzyme). During the protein synthesis, the molecular recognition between tRNA and messenger RNA is nonenzymatic, and it is governed by the Watson–Crick base-pairing scheme. However, the recognition of tRNA with the activated amino acid is enzymatic, and it is facilitated by tRNA synthetase. The connection between these two forms of recognition is evident through the intervening sequence of tRNA bases (Figure 3.25).

In the case of nucleic-acid synthesis, one enzyme does multiple jobs. On the one hand, it recognizes a template, somewhat distorts it to generate a single strand that can be copied, and reads the message written in it sequentially. The enzyme also binds substrate nucleotides at a different site and performs the catalytic synthesis. However, this synthesis has to stop at some point in time. In order to terminate the process, the enzyme recognizes some altered sequence of RNA or new protein molecules. The regulation of the synthesis and editing activities is governed by a separate set of protein–protein recognition.

The control of transcription paves the way for various finer aspects of molecular recognition. Several small molecules such as modified nucleotides and second messengers bind RNA polymerase and regulate the expression of certain genes. However, the well-defined mechanism of the control has not yet been elucidated. How the binding of a small molecule to a large surface of RNA polymerase can alter the specific promoter recognition ability is an interesting question. When antibiotics bind to RNA polymerase or the ribosome, changes in the binding domain may give rise to antibiotic tolerance. However, these changes take place due to the mutation in DNA. Small-nucleotide binding to RNA polymerase is dependent on the environment and, thus, cannot be controlled by mutation.

3.7 HORMONES AND RECEPTORS

Hormones: In multicellular organisms, growth, differentiation, metabolism, sexual differentiation, behavior, and many more complex processes are regulated by the communication between cells. Cells do not live in isolation. Cell-to-cell contact can be direct if the cells are adjacent. Cells frequently contact other cells over short and long distances by secreting signaling molecules or chemical messengers. The signaling molecule secreted by one cell is recognized by other target cells, and it induces a specific response in the target cells. An enormous number of such chemical messengers are broadly classified as cytokines, growth factors, and hormones. Cytokines and growth factors are secreted by a variety of cells, and they play a predominant role in immune response.

Hormones are secreted by endocrine glands directly into the blood. The blood carries them to the site of action, which may be anywhere in the body. Hormones can be proteins, peptides, or lipids. The major endocrine glands are pituitary, pineal, thymus, thyroid, adrenal, and pancreatic.

Over- or underproduction of hormones by any of the glands can cause profound disease conditions.

Receptors: Intercellular communication that is mediated through hormones is initiated by the activation of one or more cell surface receptors; these may be located either on or inside the target tissue. Receptors are usually proteins. Most of the receptors have two active domains: the recognition domain and the coupling domain. The recognition domain recognizes the hormone, whereas the coupling domain performs the signal transduction. Receptors for steroid and thyroid hormones have more functional domains, such as the ligand-binding domain, DNA-binding domain, and regulator-binding domain. There are different types of receptors based on their functions. The location of the hormone–receptor interaction may be extracellular, cytosolic, or nuclear.

Hormone–receptor interaction: The hormone–receptor interaction is a simple bimolecular interaction that serves as a paradigm in molecular recognition. Each hormone–receptor complex performs its function through mediator proteins such as G proteins. The hormone binds to its target receptor with a high affinity depending on its structural complementarity (Figure 3.31). A hormone–receptor interaction is very specific; different receptors can recognize even structurally similar hormones, resulting in various functions. Another important feature of the hormone–receptor interaction is that it is reversible and does not alter the conformation of hormones. This feature distinguishes a hormone–receptor interaction from an enzyme–substrate interaction. On binding to the hormone, some receptors (growth hormone receptors) undergo oligomerization (Figure 3.32). At any given time, a hormone may bind to more than one receptor or a receptor may bind to multiple hormones depending

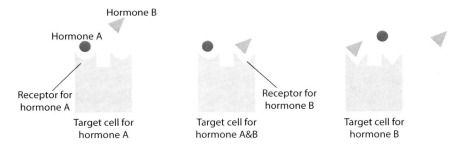

FIGURE 3.31   Hormone–receptor interaction.

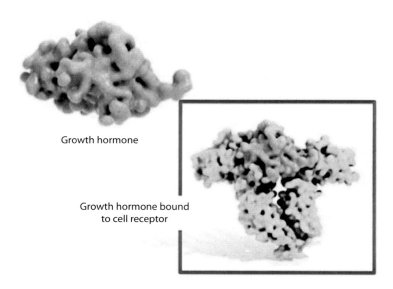

Growth hormone

Growth hormone bound
to cell receptor

FIGURE 3.32   (**See color insert.**) Dimerization of growth hormone receptor on binding to hormone. (http://ghr.nlm.nih.gov/handbook/illustrations/gh.)

on the requirement. Hydrophobic interactions and hydrogen bonds are involved in the short-range interaction between hormones and receptors. When nonpolar molecules do not protect the hydrophobic residues, they can have an effect over a long distance.

3.8 ANTIGEN–ANTIBODY RECOGNITION

The antigen–antibody complex is an excellent example for elucidating the mechanism of molecular recognition between two proteins. It is, in fact, one of the strongest noncovalent interactions known in biology. Immunoglobulins or antibodies are components of the immune system that constitute a versatile defense system of the body against foreign antigens. The main purpose of antibodies is to specifically bind pathogens or antigens, thereby marking them for the immune system. Antibodies are known to be associated with specific types of cells, are present on the B-cell membrane, and are secreted by plasma cells. B-cells are lymphocytes that are formed in the bone marrow. These cells have a B-cell receptor that binds to a specific antigen. B-cells differentiate into plasma cells, which specialize in antibody secretion. In the 1950s and 1960s, the prototype structure of the immunoglobulin (Ig) molecule was elucidated by Rodney Porter and Gerald Edelman. These experiments were considered of such great significance that the two investigators shared the Nobel Prize in 1972.

3.8.1 Structure of Antibodies

All antibodies share a basic structure. Antibodies are large molecules possessing a molecular weight of about 150 kDa. Each monomer of antibody is composed of four polypeptide chains: two "heavy" chains of about 50-kDa molecular weight and two "light" chains (lambda or kappa type) of about 25-kDa size, thus forming a tetrameric quaternary structure (Figure 3.33). A basic principle of molecular recognition is involved here. The type of heavy chain determines the immunoglobulin isotype, such as IgG, IgA, IgD, IgE, and IgM with Gamma, Alpha, Delta, Epsilon, and Mu chains, respectively. IgG, IgD, and IgE are monomeric in nature; IgM is pentameric; and IgA is monomeric, dimeric, and trimeric. Each monomeric unit has one Y-shaped unit and the units join, resulting in a multimeric form. One monomer binds to 2 antigens, hence pentameric IgM has 10 antigen-binding sites. Cartoon representation of antibody classes is shown in Figure 3.34. These two identical heavy (H) polypeptide chains and two identical light (L) chains are covalently bonded via interchain disulfide (S-S) linkages between cysteine residues. Therefore, this is a covalent recognition.

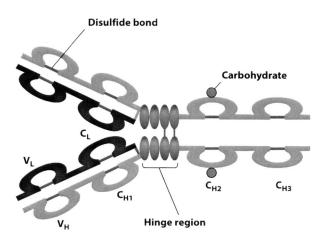

FIGURE 3.33 Structure of IgG type of antibody. V_H and V_L are the variable domains of heavy and light chains. C_H and C_L are constant domains of heavy and light chains. (http://images.slideplayer.com/27/9132209/slides/slide_7.jpg.)

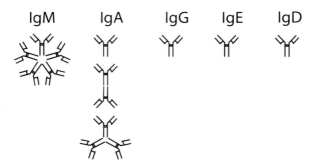

FIGURE 3.34 Antibody isotypes. (http://www.fragmenthealth.com/galery/immun-oglobulins.jpg.)

The number of interchain disulfide bonds varies among different immuno-globulin molecules. Intrachain disulfide bonds are also present within each of the polypeptide chains. The Y shape of the antibody is formed by the two equal halves of the tetramer. The region at which the arms of the antibody molecule form a Y is called the *hinge region*, because there is some flexibility at this point. The fragment crystallizable (*Fc*) region of the antibody bears a highly conserved N-glycosylation site. This region interacts with cell surface receptors called *Fc receptors*, thereby activating the immune system.

3.8.2 Constant Domain

The heavy and light chains are composed of two regions: constant (C) and variable (V) regions. Constant regions are designated as C_H and C_L and constitute a large part of the overall molecule that stabilizes the antibody molecule. Constant regions have essentially the same amino-acid sequence in all antibodies of the same class. The constant region of each heavy chain has three homologous regions: C_{H1}, C_{H2}, and C_{H3}. This region determines the class of the antibody (IgG, IgA, IgM, IgD, and IgE) and does not vary within a given class. The constant region does not interact with antigens but binds to the cell surface receptors and mediates the activation of the immune system.

3.8.3 Variable Domain

The variable domain is only a small part of the entire antibody molecule. About 100 amino-acid-long amino-terminal ends of both light and heavy chains constitute the variable region denoted as V_L and V_H, respectively. The host can produce millions of different amino-acid sequences in the variable region that are specific to different antigens with the same Y-type structure; thus, different antibodies can specifically bind to unique epitopes. The variable region of each chain contains three highly variable regions called *hypervariable* regions; these are denoted as CDR1, CDR2, and CDR3 (CDR, complementarity-determining region). These regions serve to recognize and bind specifically to antigens; they are termed *paratopes*. CDRs are separated by four invariant regions called *framework regions*; these are designated as FR1, FR2, FR3, and FR4. The role of these regions is to stabilize the spatial structure of the hypervariable domain.

3.8.4 Antigen

An antigen is a foreign substance that stimulates an immune response when introduced into the body. Antigens can be bacteria, viruses, protozoa, foods, pollen grains, serum components, cells, and tissues of various species, including humans. The immune system, especially lymphocytes, recognizes distinct regions of antigens through protruding regions on the large antigenic or immunogenic molecule; these are known as *epitopes or antigenic determinants*. These regions are recognized by antigen-specific receptors or antibodies.

When antigens are combined with adjuvants, they enhance the immune response and produce a higher titer of antibodies. Adjuvants such as aluminum hydroxide and parafilm are agents that when combined with antigen give long-lasting immunity; this is due to the slow release of antigens. They are mostly used in the case of vaccine preparation and for the production of antibodies in immunized animals.

3.8.5 Antigen–Antibody Recognition

In this molecular recognition, an epitope region of the antigen is recognized by the paratope of the antibodies (Figure 3.35). This is confirmed by x-ray crystallography of various antigen–antibody (Ag–Ab) complexes. The Ag–Ab binding depends on the distance between antigen and antibody, shape, complementarities of antigen and antibody, antibody affinity, noncovalent bonds, and other intermolecular interactions (Figure 3.36).

Ag–Ab binding could be due to weak or strong interactions. In the case of weak interactions, the antibody binds to the antigen by noncovalent interactions; these include hydrogen bonds, ionic bonds, hydrophobic interactions, and van der Walls interactions. Since these interactions are relatively weak, a large number of these interactions are needed for strong antigen–antibody interactions, which are effective only over a

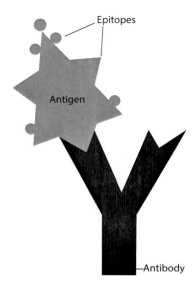

FIGURE 3.35   Antigen and antibody binding.

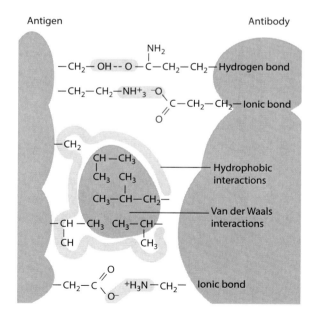

FIGURE 3.36 Intermolecular interactions in antigen–antibody binding. (Adapted and modified from http://bbs.pranovo.com/.)

short distance, namely, 1 Å. For a strong binding and to elicit an effective immune response, the epitope and its binding region of the antibody should have a complementary shape. This juxtaposes the interacting groups in close proximity. The size of the epitope should be lesser than the size of the antibody's binding region. Protruding epitopes fit into the depressions of the paratope. Around 15–22 amino acids of the epitope region interact with the antibody through 75–120 hydrogen bonds as well as by ionic and hydrophobic interactions. The surface topography of the antigen-binding surface of the antibody can vary considerably according to the size and shape of the antigen. For large protein antigens, the contact surface of the antibody is generally planar; whereas for small antigens, it is more concave such that the antigen surface is hidden. Antibody binding to medium-sized antigens has a grooved surface with distinct cavities in its binding region. In some cases, the binding of Ag–Ab induces a conformational change in both antigen and antibody.

Antibody affinity is defined as the combined strength of noncovalent interactions between a single epitope and a single paratope. Low-affinity antibodies weakly bind to antigens and can easily disassociate from them. High-affinity antibodies bind to antigens more tightly and remain bound for a longer time. In some cases, antigen–antibody association is observed for 24 hours. Ag–Ab interaction can be monomeric in nature, as in the case of IgG, IgE, and IgD antibodies and multimeric, as displayed in IgM and IgA antibodies. Antigen–antibody interfaces are rich in aromatic residues, especially tyrosine and tryptophan, but devoid of charged residues such as aspartate, glutamate, and lysine. It appears that the aromatic stacking interaction plays an important role in overall recognition. This interface is also enriched with arginine. The surface area of the interaction is about 600–900 Å.

The complex between an antibody binding to an antigen is shown in Figure 3.37. Residues from the antibody interact through H-bonding, electrostatic reactions, and van der Waals interactions.

3.9 BLOOD GROUPING

Another classic example of molecular recognition is the recognition of cell surface glycoprotein by the antibodies of the immune system. Variations in polysaccharides at the cell surface of red blood cells define the blood groups of an individual; these variations exhibit remarkable specificity. However, the difference in polysaccharides is controlled by the genetic

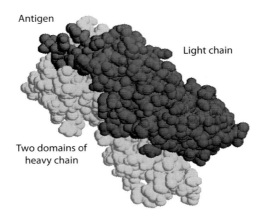
Antigen

Light chain

Two domains of
heavy chain

FIGURE 3.37   (**See color insert.**) Binding of HIV antibody to HIV antigen.
(http://courses.washington.edu/.)

system of the individual and the degree of difference among them is very
subtle.

In recent times, 33 blood grouping systems have been approved.
Among these, the ABO blood group system has the most clinical significance in the case of blood transfusion and organ transplantation.
The ABO blood group is determined by the presence of A and B antigens on the surface of red blood cells (RBCs) and of anti-A and anti-B
antibodies in the plasma. These antibodies are usually IgM type and
are not present in newborns. They appear in the first year of life, maybe
due to exposure to different food and environmental antigens that are
similar to A and B antigens. In 1930, Karl Landsteiner was awarded
the Nobel Prize for the discovery of ABO blood grouping. Without the
knowledge of Landsteiner's work, Jan Jansky independently worked on
blood grouping. He was renowned for his classification of blood into
four types, such as A, B, O, and AB.

The second most important blood group system is the rhesus system
or Rh system. It is classified based on the presence or absence of the D
antigen or Rh factor on the RBC surface and is denoted as Rh⁺ and Rh⁻
type. An Rh-negative individual produces anti-Rh antibodies only when
exposed to the Rh antigen. These Rh antibodies are of IgG type, hence
they can cross the placenta. If a pregnant mother is Rh negative and the
fetus is Rh positive, the mother's immune system produces Rh antibodies.

TABLE 3.1   Common Blood Grouping Systems

Blood Type	RBC Surface Antigens	Antibodies in Plasma	Rh or D Antigen
A positive	A antigen	Anti-B antibody	Present
A negative	A antigen	Anti-B antibody	Absent
B positive	B antigen	Anti-A antibody	Present
B negative	B antigen	Anti-A antibody	Absent
AB positive	A and B antigens	No antibody	Present
AB negative	A and B antigens	No antibody	Absent
O positive	No antigen	Anti-A and anti-B antibodies	Present
O negative	No antigen	Anti-A and anti-B antibodies	Absent

These antibodies do not cause problems during the first delivery, because the baby is born before many of these antibodies develop. However, in the case of a second delivery, if the fetus is again Rh positive, the mother's Rh antibodies pass through the placenta and attack the fetus' RBC. This Rh incompatibility leads to hemolytic anemia in the baby, which may be a fatal condition. To prevent this, Rh antibodies are injected into the mother who is carrying an Rh-positive fetus.

Table 3.1 summarizes the ABO and Rh blood grouping systems.

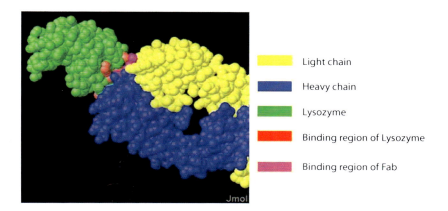

FIGURE 1.8 Space-filling model of lysozyme binding to its antibody. (http://mcdb-webarchive.mcdb.ucsb.edu/sears/immunology/Antibody-Antigen/hel-fab-f.htm.)

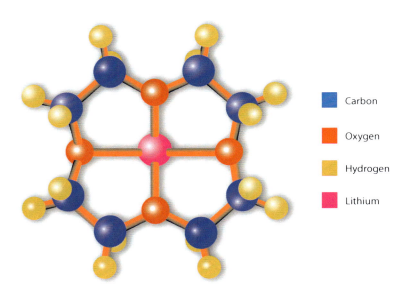

FIGURE 2.3 12-Crown-4 lithium-ion complex. (Adapted and modified from http://www.chem.ucla.edu/harding/crownethers.html.)

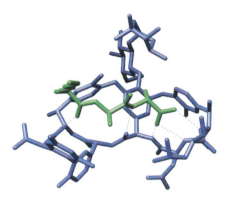

FIGURE 2.4   Crystal structure of L-Lys D-Ala–D-Ala with vancomycin (green represents the antibiotic, and blue represents the tripeptide). (From Knox, J.R., and Pratt, R.F., *Antimicrob Agents Chemother,* 1342–1347, 1990, https://en.wikipedia.org/wiki/Vancomycin.)

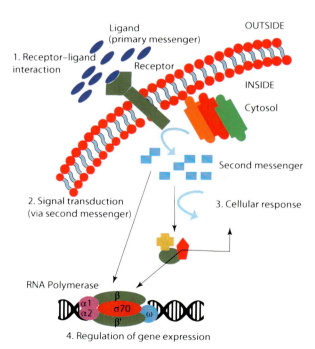

FIGURE 2.7   Schematic representation of a signal transduction pathway in bacteria. A primary signal from outside the cell is transferred inside an effector molecule through a series of recognition processes. (Adapted and modified from Bharati, B.K., and Chatterji, D., *Curr. Sci.,* 105(5), 643, 2013.)

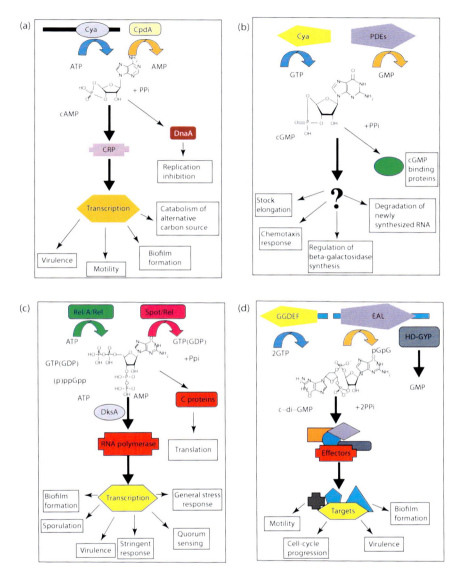

FIGURE 2.9   A schematic representation of second messenger signaling modules in bacteria. Cyclic AMP, cyclic GMP, ppGpp, and c-di-GMP signaling modules are shown in (a)–(d), respectively. (Adapted and modified from Pesavento, C., and Hengge, R., *Curr. Opin. Micrbiol.*, 12(2), 170–176, 2009.)

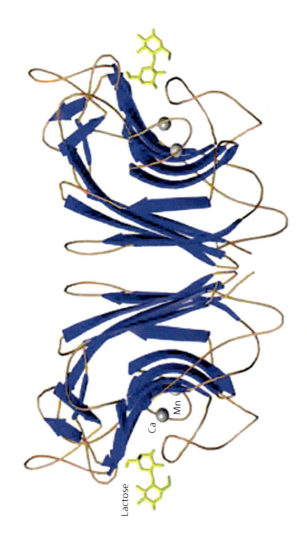

Lactose

Ca

Mn

FIGURE 2.10 Structure of an oligosaccharide bound to a part of the lectin plant. (http://mbu.iisc.ernet.in/~kslab/.)

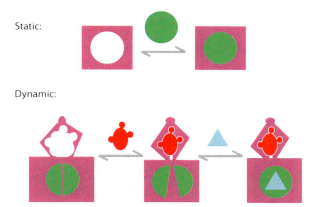

Static:

Dynamic:

FIGURE 3.1   Static and dynamic binding

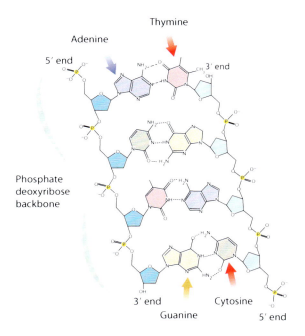

Thymine

Adenine

5′ end

3′ end

Phosphate
deoxyribose
backbone

3′ end   Cytosine

Guanine   5′ end

FIGURE 3.5   Nucleic-acid base pairs. (Adapted and modified from https://en.wikipedia.org/wiki/File:DNA_chemical_structure.svg.)

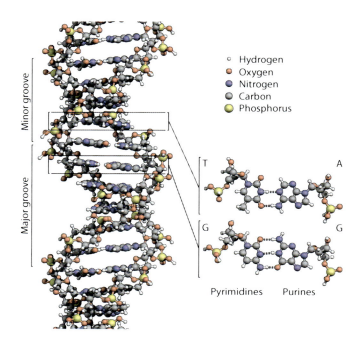

FIGURE 3.6   Right-handed B DNA. (https://upload.wikimedia.org/wikipedia/commons/4/4c/DNA_Structure%2BKey%2BLabelled.pn_NoBB.png.)

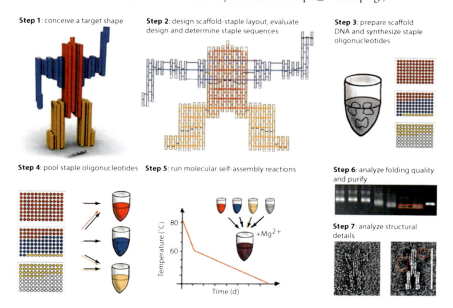

FIGURE 3.7   DNA Origami. (Carlos, C.E. et al., *Nature Methods*, 8(3), 2011.)

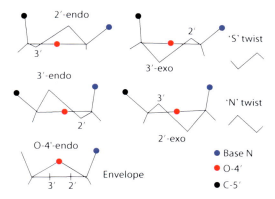

FIGURE 3.8   Puckered structures of ribose sugars. (Adapted and modified from fbio.uh.cu.)

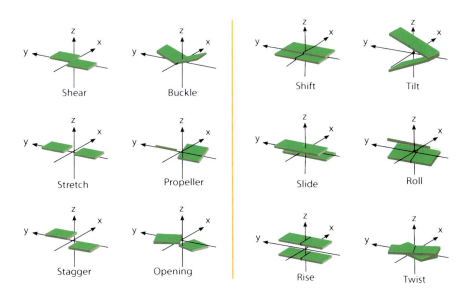

FIGURE 3.11   Base pairs in altered conformation. (Adapted and modified from http://x3dna.org/.)

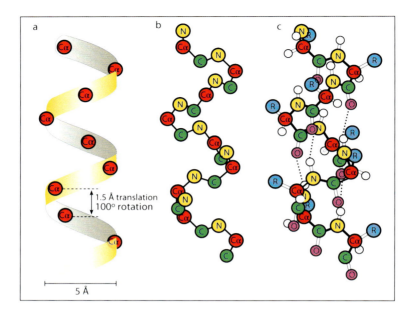

FIGURE 3.14   Alpha-helix. (Adapted and modified from http://cmgm.stanford.edu/.)

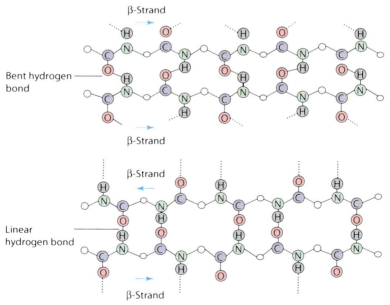

FIGURE 3.16   Parallel strands with bent H-bonds and antiparallel strands with linear hydrogen bonds.

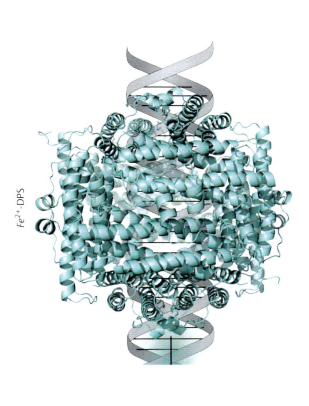

Fe^{2+}-DPS

FIGURE 3.21 Iron-bound DPS. (Arnold, A.R., and Barton, J.K., *J. Am. Chem. Soc.*, 135(42), 15726–15729, 2013.)

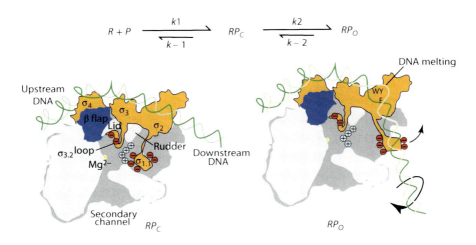

FIGURE 3.23   Binding of *E. coli* RNA polymerase with T7A1 promoter. (Murakami, K.S., and Darst, S.A., *Curr Opin Struct Biol* 13(1), 31–9, 2003.)

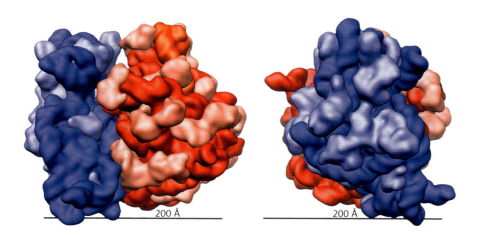

FIGURE 3.24   Association of 50S (RED) and 30S (BLUE) subunits to form 70S bacterial ribosome. (https://upload.wikimedia.org/wikipedia/commons/4/42/ Ribosome_shape.png.)

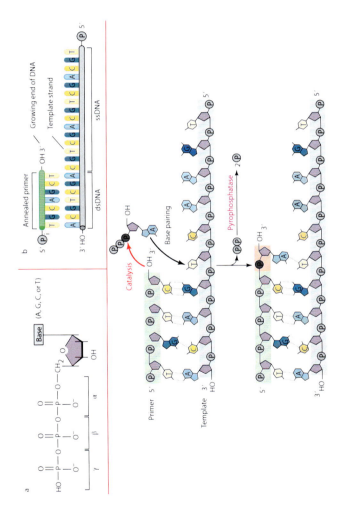

FIGURE 3.27 Chemistry of DNA synthesis. (Pearson Education, Inc., http://www.slideshare.net/narayanprahlad/dna-replication-and-enzymes-involved-in-dna-replication.)

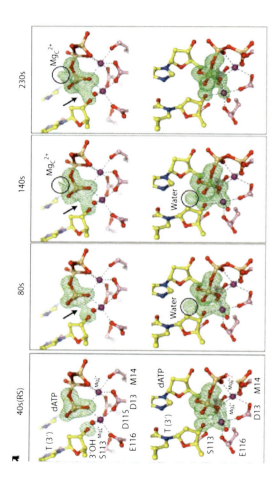

FIGURE 3.28 Reaction time course of DNA polymerase. The reaction process is monitored by the electron density corresponding to the new chemical bond in the map of the reactant state (RS) compared with the refined GS (ground state). The emerging densities are pointed out or circled. RS is the reactant state structure where there is no sign of bond formation. Two unexpected spheres of electron density appear in the course of new bond formation. The first is within hydrogen bonding distance of the 3′-OH. Its electron density peaks at 80s and drastically declines after 140s when the PS (product state) becomes prominent. Water is hydrogen bonded with the 3′-OH, the O4′ of dATP, and water in the 80s RS structure. The second emerging sphere of electron density is very close to the α-phosphate on the opposite side of the A- and B-site Mg2+ ions. It starts appearing at 140s and intensifies with the reaction time. Without an A-site divalent cation, the 3′-OH of the primer strand shifts away from the dATP and forms hydrogen bonds with the side-chains of S113 and D115. The active-site carboxylates D13 and E116 adopt different rotamer conformations. (Reprinted by permission from Macmillan Publishers Ltd. Nature, Nakamura, T. et al., Watching DNA polymerase η make a phosphodiester bond, 487, 196–201, copyright 2012.)

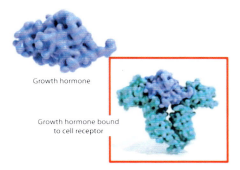

Growth hormone

Growth hormone bound
to cell receptor

FIGURE 3.32   Dimerization of growth hormone receptor on binding to hormone. (http://ghr.nlm.nih.gov/handbook/illustrations/gh.)

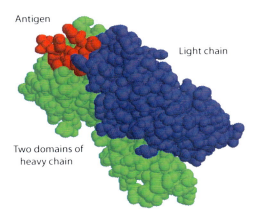

Antigen

Light chain

Two domains of
heavy chain

FIGURE 3.37   Binding of HIV antibody to HIV antigen. (*Source:* http://courses. washington.edu/.)

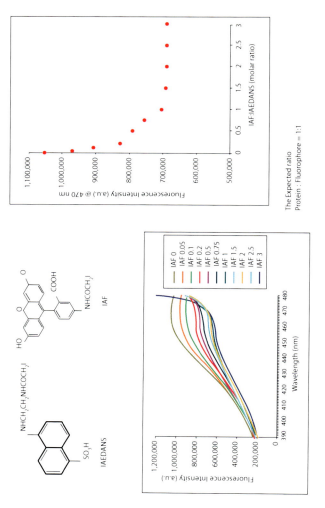

FIGURE 4.7 Labeling of protein with N-(Iodoacetaminoethyl)-1-naphthylamine-5-sulfonic acid (IAEDANS) and 5-Iodoacetamidofluorescein (IAF).

FIGURE 5.2 Charge transfer–induced folding of a donor–acceptor oligomer. (Adapted and modified from Macmillan Publishers Ltd., Nature, Lokey, R.S., and Iverson, B.L., Synthetic molecules that fold into a pleated secondary structure in solution, 375, 303, copyright 1995.)

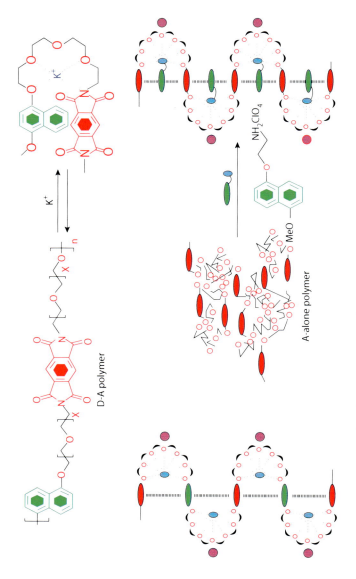

FIGURE 5.3   Charge transfer–induced folding of truly flexible donor–acceptor polymer and acceptor-alone polymer. (Adopted and modified from Ghosh, S., and Ramakrishnan, S., *Angew Chem. Int. Ed.*, 43, 3264, 2004; 44, 5441, 2005.)

Methods to Follow Molecular Recognition

In this chapter, we learn about the different ways in which one can study the strength of molecular recognition by using various tools and techniques. Why is it important to estimate the binding strength between two recognition partners? The stability of the complex formed is intimately related to the strength of the interaction. Association and dissociation of the noncovalent complex measure its stability, and there are different ways in which one can estimate this value. The progressive change in any measurable property of the macromolecule owing to its interaction with a specific partner can be monitored using various methods, some of which are addressed in this chapter.

In a typical scenario, when two molecules are mixed together and if they have an affinity for each other, they will associate. Thus, at a given point of time, the solution will contain complex and unreacted species.

To quantitate the strength of molecular recognition, it is necessary to separate the complex from its constituents. Most often, difficulties arise when the complexes are unstable as they are weak noncovalent interactions. The most stable noncovalent interaction is estimated to have the dissociation constant K_d in picomolar range for simple stoichiometric binding. (For a 1:1 binding between a ligand and a receptor, K_d is defined as that concentration of ligand at which the receptor is 50% saturated with the ligand). However, antigen–antibody recognition and the streptavidin–biotin interaction are more stable. Biotin is a soluble vitamin analog that binds the tetrameric protein avidin or streptavidin with the dissociation constant K_d in the order of 10^{-15} M. The problem is further compounded if one is trying to monitor the monomeric recognition. The most reliable method that is used for following the monomer–monomer recognition probably involves analyzing the crystal structure of the complex; this accurately displays hydrogen and other types of bonding between monomers. The method is not easy to follow and is esoteric in nature. Specific interactions between amino acids and nucleic acid bases, and the recognition between two complementary bases, are documented with x-ray analysis.

Nuclear magnetic resonance (NMR) has been extensively used to follow hydrogen bonding between Watson–Crick base pairs at a low temperature and at times in nonaqueous medium. Hydrogen bond formation leads to a downfield shift of protons in a certain area of the spectrum; this is easily discernable, as the protons are aromatic in nature. The concentration-dependent shift helps in estimating the strength of the interaction.

It is easy to quantitate the entire process of molecular recognition when a polymer recognizes a monomer unit of various sizes. The same principle holds good when the recognizing partners are of different sizes, even when both are macromolecules. In general, the complex formation is monitored based on differences in the following: (1) size, (2) charge, (3) affinities for metal ions or receptors, and (4) hydrophobicity.

4.1 SIZE-EXCLUSION CHROMATOGRAPHY

According to this method, molecules in solution are separated by their size or molecular weight. This method is usually applied to large molecules or macromolecular complexes such as proteins and industrial polymers. To separate constituents, the complex is passed through a carefully selected matrix. The larger molecules are eluted first in the void volume followed by the smaller ones. It is easy to approximate the void volume on the basis of the column volume. In a 30-ml column, the void volume is 10 ml, or always one-third.

Thus, the elution of the smaller molecule in the void volume is a measure of the complex formation between the two interacting partner. Any change in the incorporation of the monomer in the void volume as

FIGURE 4.1   Electrophoretic mobility shift assay. Lane 1, free DNA. Lane 2, DNA–protein complex generating a high-molecular-weight species that cannot move down from the well.

a function of its concentration can be exploited to estimate the binding parameters, at least for one or multiple independent recognition sites.

A similar principle applies when the complex is separated through polyacrylamide or agarose gels. The mobility of the macromolecular complexes through these gels varies according to size, provided they have a uniform distribution of charge in the backbone. For example, DNA are negatively charged by nature; thus, they move toward the positive electrode. However, to gain a uniform distribution of charge in proteins, they need to be treated with detergent. Smaller molecular-weight species will move faster than larger ones. This technique has been extensively used. Agarose gels are well suited for the separation of DNA base pairs that are a few hundred or above in length, but polyacrylamide gels can resolve base pairs of less than 100 in numbers.

The method mentioned above monitors DNA–protein interactions and is known as an electrophoretic mobility shift assay (EMSA). When a sequence-specific DNA–protein complex is formed, the molecular weight of the species is higher than that of the free DNA. Because of its high molecular weight, the DNA–protein complex is expected to move slower on a gel than the free DNA. On the basis of the relative distribution of DNA between the free and the complex form, we measure the strength of binding between the two. One such DNA–protein complex identified by EMSA is shown in Figure 4.1. In practice, independent multiple binding sites on DNA give rise to multiple retarded complex bands.

4.2 SURFACE PLASMON RESONANCE

This dynamic technique is used to monitor the ligand–receptor interaction by following the change in size, at least indirectly. The bait molecule is immobilized on a surface, and the refractive index is measured. One can detect the light from total internal reflection falling on the detector after bouncing off the surface. Subsequently, the prey is passed through the surface and the formation of the complex results in change in size, as a consequence change in the refractive index. This method is highly popular for estimating specific protein–protein or DNA–protein interactions. Even small-molecular recognition and drug–DNA binding are followed by surface plasmon resonance (SPR). The major limitation, however, is to immobilize the bait on a solid surface by covalent modification. Different shapes and sizes of the curve indicate the nature of recognition between the two species.

Figure 4.2 shows binding and dissociation curves of a small-molecule ligand, in this case 3-formyl rifampicin (Figure 4.3), which can be easily

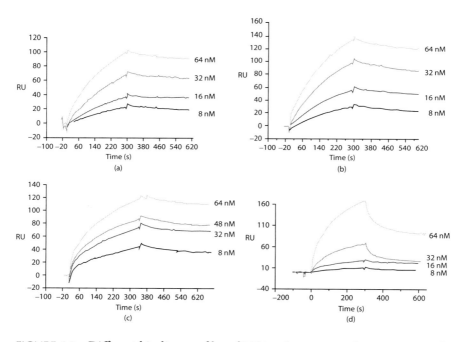

FIGURE 4.2 Different binding profiles of RNA polymerase and its mutants with 3-formyl rifampicin. Here, the rifampicin derivative is fixed on the surface and RNA polymerase of a different concentration is flowing as a function of time. RU, response units.

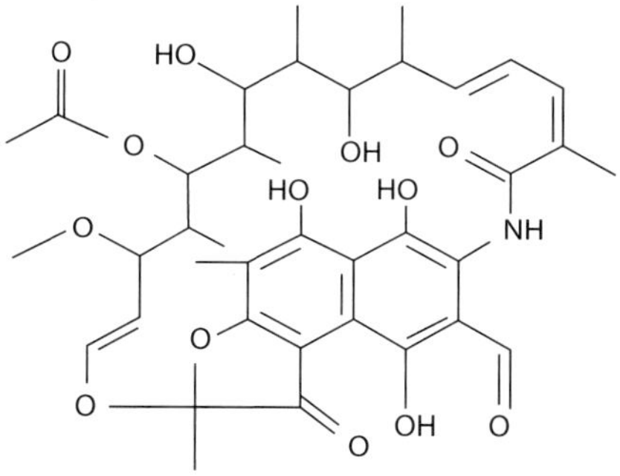

FIGURE 4.3 3-Formyl rifampicin. (Adapted and modified from http://www.chemicalbook.com/.)

immobilized on a solid surface. When bacterial RNA polymerase is passed over it at a different concentration, it shows binding curves of different shapes and the binding is strong with estimated Kd, which is in nanomolar range. Different forms of the RNA polymerase enzyme were used here in four different cases; each shows different binding profile. Rifampicin alone cannot be immobilized and thus the modification is necessary. One needs to appreciate that such chemical modification is at times necessary to improve the immobilization technique. However, it is necessary to ensure before the commencement of the experiment that the modified ligand behaves similar to the unmodified one in the biological assay.

Rifampicin inhibits bacterial transcription by binding to RNA polymerase, but it does not bind to eukaryotic RNA polymerase. Therefore, it serves as a very good drug against bacterial infection.

4.3 AFFINITY CHROMATOGRAPHY

If one of the interacting partners has an affinity for small molecule–like metal ions, the other partner can also be trapped by the metal–ligand complex, even if it does not have any metal binding surface. The affinity between the two macromolecules is the deciding factor for the stability of the complex, and this can be monitored as a function of the concentration of the second ligand. This method has been extensively used to purify proteins with an abundance of histidine amino-acid residues at one of their terminals. Since Ni(II) has strong affinity for the imidazole ring in histidine, the protein will bind the Ni(II) matrix, and the Ni(II)-bound protein can be used as bait to pick up its recognition partner. A specific receptor bound to a matrix or an antibody affinity matrix works on the same principle. An antibody generated for one specific partner can bind either the protein alone or its complex, where the other interacting partner of the complex has no affinity for the antibody. Co-immunoprecipitation and receptor-bound affinity chromatography are routinely used in this manner.

Protein complex bound to Ni(II) can be eluted out by a competing ligand such as imidazole in excess. A change in pH can also be used to release the protein from the matrix. As one of the nitrogens in imidazole has low pI and it coordinates with Ni(II), the change in pH results in protonation of the nitrogen and release of the protein. Ni(II)-based affinity chromatography poses a great advantage. Even from a crude mixture of proteins and DNA, one can recognize the partner if the other partner is immobilized through Ni(II) chemistry. Antibody-based co-immunoprecipitation is also very useful to detect molecular recognition between two proteins. This is possible due to the strength of the antigen–antibody interaction.

4.4 ION-EXCHANGE CHROMATOGRAPHY

A complex that is formed between two macromolecules due to an ionic interaction can be dissociated by changing the ionic strength of the media. Thus, the molecules can be separated either by ion-exchange chromatography or by the addition of salt or polyanions to the complex. Charge-based affinity chromatography is also well known. When attached to a cellulose matrix, polyanions such as heparin or phosphate can act as a DNA mimic and are, therefore, used for specific recognition of DNA-binding proteins. These proteins can, therefore, be separated from a mixture of proteins by passing them through this matrix. The strength of the interaction can also be evaluated based on the concentration-dependent dissociation of the complex in the presence of a salt.

It was mentioned earlier that NMR is a useful technique that is utilized for molecular recognition. Similarly, other spectroscopic methods can also detect the complex formation. Either the change of a signal due to the complex formation or the appearance of a new signal is considered a tool that is utilized for the interaction. Different spectroscopic techniques such as ultraviolet spectroscopy, infrared spectroscopy, or circular dichroism (CD) are very useful at times.

If co-factors are present in protein-like small organic molecules or metal ions, they can act as markers as well. Nucleic acid–protein recognition has been extensively studied owing to its central role in biology, and it is also amenable to study by a variety of techniques. However, this is not the case with protein–carbohydrate recognition. Affinity chromatography-based methods are in greater demand in such cases. It is not easy for membrane protein–receptor recognition to utilize routine techniques.

4.5 HYDROPHOBICITY

The presence of high salt dissociates an ionic partner and helps to a great extent in identifying the macromolecules involved in molecular recognition. On the other hand, when the recognition takes place exclusively by a hydrophobic interaction, high salt promotes the stability of such species. Thus, if one needs to separate the molecules, a different approach must be envisaged. Often, a small molecule elutant that has more affinity for one of the partners is used. Hydrophobic solvents also dissociate the complex and can help in estimating the strength of the interaction.

When molecules have both hydrophilic and hydrophobic surfaces, they are known as *amphipathic molecules.* These molecules utilize both the surfaces to recognize their respective partners. Several DNA-binding proteins exhibit these properties by which hydrophobic surfaces of amino-acid side chains are lined up together. To measure the strength of binding between hydrophobic molecules or between amphipathic molecules, a change in surface pressure is often employed. This is known as the Langmuir–Blodgett (LB) technique.

In this method, a trough is used for filling water; a minimal amount of an amphipathic molecule dissolved in a suitable solvent is layered on top of this, so that it generates a monolayer at the air–water interface. The amphipathic molecule, which is most often a divalent metal salt of a fatty acid, will spread such that the hydrophilic part [Ni(II) complexed with carboxylate] stays within the bulk aqueous medium, whereas the hydrophobic part (fatty acid chain) protrudes at the air–water interface. When such a layer is compressed with a movable mechanical device (a barrier) attached to a computer on the outside, the surface pressure increases, as shown in Figure 4.4.

There are three distinct phases of surface pressure during compression. At the beginning, the molecules at the interface are randomly oriented and this is called the *gas phase,* which possesses maximum entropy. However, when the barrier moves toward compressing the surface, molecules get increasingly ordered and, ultimately, saturation in surface

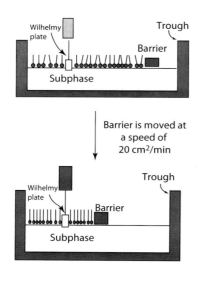

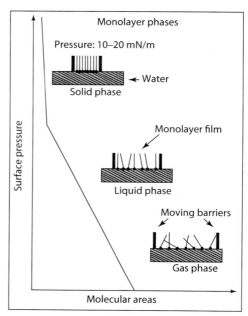

FIGURE 4.4   Langmuir–Blodgett technique.

pressure is attained. This is known as a *solid phase* or a *monolayer*. Any further compression of the monolayer will result in the collapse and piling up of surface molecules over each other.

Interestingly, one can move the barrier both back and forth to generate a hysteresis curve. It is also possible to insert a surface with a hydrophilic or hydrophobic coating into the trough to transfer the monolayer. These glass surfaces are called Wilhelmy plates. The transferred monolayer can then be studied by various means.

Macromolecular recognition between two species will result in higher molecular mass. This was demonstrated by SPR—another surface-related technique—that the change in mass results in a change in refractive index, which is an important measurable parameter for recognition. In the case of Langmuir, the change in surface pressure due to the increase in mass is employed in order to follow the macromolecular recognition. Thus, it is necessary to have an amphipathic molecule at the surface. Tethering a macromolecule to a nonreactive buoy floating on the surface can also generate the monolayer. An example of the Ni(II) complex of fatty acid that generates a monolayer was

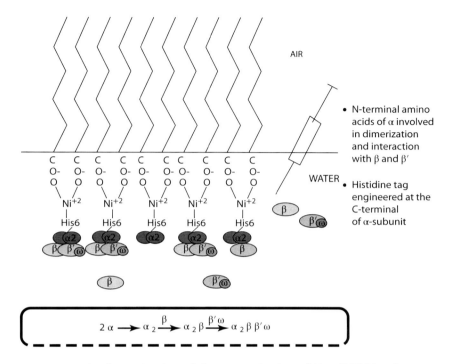

FIGURE 4.5 A schematic view of the reconstitution of *E. coli* RNA polymerase at Langmuir monolayer.

given earlier. This was not without a reason, as the Ni(II) in the aqueous medium can recognize a protein with a histidine tag and the protein can be suspended from the monolayer as well. This can be used to assemble either a protein–protein complex or a protein–DNA complex. The additional advantage of the LB technique is that the monolayer that is transferred to the Wilhelmy plate can be studied by microscopy or a spectroscopic method such as Fourier transform (FT)-infrared spectroscopy. In Figure 4.5, the assembly of a large enzyme complex such as RNA polymerase is demonstrated; this sequential assembly can be achieved when only one subunit is attached to the buoy.

4.6 SPECTROSCOPIC TOOLS THAT UTILIZE MOLECULAR RECOGNITION

We mentioned earlier in this section that NMR experiments are very useful while utilizing DNA base pair recognition. In the case of larger complexes, NMR analysis can still be conducted; however, the assignment of peaks arising due to certain specific groups is not an easy task and requires adequate experience. Multidimensional NMR and FT-NMR are used to increase either the resolution or the signal-to-noise ratio, respectively.

A change in the UV–visible spectrum—caused by recognition—can yield qualitative results, but a quantitative estimation of the binding constant is very approximate. However, different researchers have utilized these methods to investigate the stoichiometry of the complex by studying the shift in the pattern that is caused by the interaction of molecules and the generation of isosbestic points, as shown in Figure 4.6.

An *isosbestic point* is defined as the wavelength (in this case, at 510 nm) at which the total absorption of the complex does not change owing to chemical transformation. When A reacts with B, in the specific example depicted earlier, where A indicates DNA and B indicates ethidium bromide, the visible spectral peak of the dye shifts due to the DNA interaction with the dye. If all the resulting spectra have the same absorption at a

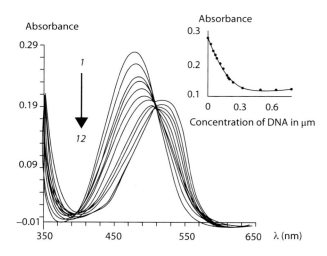

FIGURE 4.6   Changes in the absorption spectra of ethidium bromide dye at the visible region due to the successive addition of DNA. Ethidium bromide intercalates DNA, resulting in a shift of the absorption maxima. It also gives pink fluorescence due to intercalation. (Baranovsky, S.F., Bolotin, P.A., Evstigneev, M.P., Chernyshev, D.N.J., *Appl. Spectrosc.*, 76(1), 132–139, 2009.)

particular wavelength, this is known as an *isosbestic point*. This indicates that there is only one type of complex.

Fluorescence spectroscopy, on the other hand, is more useful, as no DNA bases are fluorescent and three amino acids exhibit weak fluorescence signals. Out of the three (W, Y, and F), only W has good quantum yield (0.12–0.14) and it is also studied most often. Thus, in the event of DNA–protein recognition, the nature of the interaction can be followed as well as quantitated by measuring the signal emitted from tryptophan residue in the protein. At times, there is only one W residue or a few such residues buried inside the protein. As the fluorescence signal is sensitive to the environment, pH, charge, and depth of the residue in the protein, molecular recognition between specific partners would result in a change in the fluorescence emission of the protein.

Very often, extrinsic fluorescent probes are attached to the interacting partners and if the biological function of the macromolecule is not altered due to labeling, these probes are selected as markers. DNA-specific or amino acid–specific probes, most often cysteine residue, are available. Sugars are spectroscopically silent in nature but specific fluorescent probes attached to a particular location within the sugar molecule can make them amenable for detailed investigation.

In Table 4.1, the quantum yield (number of photons emitted divided by the number of photons absorbed) and fluorescence intensities of different probes are given.

Fluorescent spectroscopy in molecular recognition finds another important application. This method deals with the quantitation of the geometric relationship between different domains in a protein or the DNA–protein complex by measuring the distance between two sites labeled with two different fluorescent probes. This is known as Förster's resonance energy transfer (FRET). If the emission spectrum of one probe, called the *donor*, overlaps with the excitation or absorption spectrum of another probe, called the *acceptor*, and the donor–acceptor pair is placed within a 10–60 A° distance, then dipolar energy transfer would occur between them. This results in a quenching of the emission spectrum of the donor and an enhancement of the excitation or absorption spectrum of the acceptor; the degree of change can be interpreted as energy-transfer efficiency. In Box 4.1, this is shown in a figure and via a few equations in mathematical terms. A typical example of a cysteine-specific fluorescent

TABLE 4.1 Fluorescent Probes and Their Properties

Extrinsic Probe	Labeling Method	Absorption Maxima λ_{max} (nm)	Extinction Coefficient $\varepsilon_{max} \times 10^{-3}$	Emission Maxima		
				λ_{max} (nm)	Quantum Yield	τ_F (ns) Lifetime
Dansyl chloride	Covalent attachment to protein: Lys, Cys	330	3.4	510	0.1	13
Fluorescein isothiocyanate (FITC)	Covalent attachment to protein: Lys	495	42	516	0.3	4
Ethidium bromide	Noncovalent binding to nucleic acids	515	3.8	600	0.9	26.5
Proflavine	Covalent attachment to RNA 3′ ends	445	15	516	0.34	30
1,5-N-(Iodoacetaminoethyl)-1-naphthylamine-5-sulfonic acid	Covalent attachment to protein: Lys, Cys	360	6.8	480	0.5	15
Iodoacetamide	Thiol-reactive dye	511	1.15	533	0.8	3.9
Tryptophan	Intrinsic probe	280, 295	5.6	345	0.14	2.6
Tyrosine	Intrinsic probe	274	1.4	303	0.08	3.6

probe, extrinsic in nature, is shown in Figure 4.7. This probe emits light in the visible region and acts as a donor. When a suitable acceptor is added to this solution, the quenching of the fluorescence of the donor can be recorded in a stoichiometric ratio. Thus, these pairs can be used as a Förster's pair to estimate the distance between them when the protein is labeled with probe covalently.

BOX 4.1   FÖRSTER RESONANCE ENERGY TRANSFER

Förster resonance energy transfer (FRET), named after Theodor Förster, is often called fluorescence resonance energy transfer, as most often two chromophores are fluorescent in nature. However, energy is actually never transferred by fluorescence. As both these forms of transfer are abbreviated as FRET, this misrepresentation happens. Strictly speaking, FRET is not limited to fluorescence alone. This is because dipolar energy transfer can happen between any two light-sensitive chromophores that are suitably placed with respect to each other. At its electronic excited state, a donor chromophore may transfer energy to an acceptor chromophore through *dipole–dipole coupling*, which is nonradiative in nature. The efficiency of this energy transfer is inversely proportional to the sixth power of the distance between the donor and the acceptor; this is evident in the following equation:

$$E = \frac{1}{1 + (r / R_0)^6}$$

where E is defined as the quantum yield of energy-transfer transition or the fraction of energy transfer that is actually occurring per single-donor excitation, R_0 is the constant or the distance between a fixed pair consisting of a donor and an acceptor at which 50% of energy transfer takes place between both the donor and the acceptor, and "r" is the actual distance between the donor and the acceptor. Easy measurements and availability of several matching fluorophores make FRET extremely sensitive to small changes in distance. The radiation-less transition is an interesting phenomenon. Since the excited chromophore emits a virtual photon that can never be detected but is captured by the acceptor chromophore, the term "radiation-less" is used. Donor's radius of influence is much less than the wavelength of the light at which the FRET is measured. The accompanying diagram explains the process through the well-known Jablonski diagram.

(continued)

BOX 4.1 (Continued) FÖRSTER RESONANCE ENERGY TRANSFER

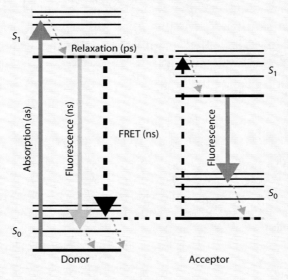

(Adopted and modified from https://upload.wikimedia.org/wikipedia/commons/thumb/c/c4/FRET_Jabolinski_Diagram.svg/2000px-FRET_Jabolinski_Diagram.svg.png.)

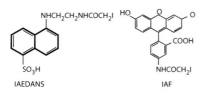

IAEDANS   IAF

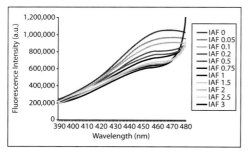

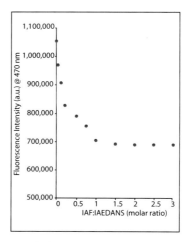

The expected ratio
Protein : Fluorophore = 1:1

FIGURE 4.7 **(See color insert.)** Labeling of protein with N-(Iodoacetaminoethyl)-1-naphthylamine-5-sulfonic acid (IAEDANS) and 5-Iodoacetamidofluorescein (IAF).

It can be seen from the equation shown in the box that the energy-transfer efficiency is distance dependent and can be monitored to follow conformational change. Thus, the change in protein conformation due to the interaction of protein with DNA, sugar, or any small molecule can be measured. Intrinsic tryptophan residue or tethered extrinsic probes could be utilized as a donor or an acceptor. All one needs is a good spectral overlap between the donor and the acceptor. The variation in energy transfer between a DNA-bound probe and a specific tryptophan within a protein as a function of time can estimate the movement of the protein over DNA. The addition of a denaturant or any other perturbation such as a shift of temperature would also result in a change in distance. Such measurements usually throw light on the biological mechanism behind recognition. Lastly, any molecule with an absorption signal but with very weak emission could also be used as an acceptor for Förster distance measurement.

All the techniques discussed earlier are much needed tools to study macromolecular recognition. Other techniques are also based on the same principles of spectroscopy as mentioned earlier. For instance, CD (circular dichroism) is an elegant spectroscopic method that monitors conformational changes caused by an interaction between two species. The roots of this optical activity–based technique lie in the fact that both nucleic acids and proteins are optically active with a distinct CD profile at different wavelengths. Three major conformational profiles of protein have signature CD spectra, as shown in Figure 4.8. Thus, one can estimate the percentage population of each species in a protein preparation and any change in them due to DNA or sugar recognition.

There are several programs available that can provide the approximate percentage of structure present in a given protein. From Figure 4.8, one can readily conclude that myoglobin is a typical α-helical protein.

Within the cell, DNA predominantly exists as right-handed B-DNA with a typical CD characteristic—a negative peak at 255 nm (Figure 4.9). However, certain synthetic sequences in DNA can be transformed to a left-handed Z-DNA form having a different CD profile, with a negative peak at 280 nm. Under extreme dehydration or in the presence of high salt, such a transition may take place. In a typical test tube experiment, as shown in Figure 4.9, a transition from B- to Z-DNA is recorded. However, the existence of left-handed DNA or Z-DNA within the cell is still controversial.

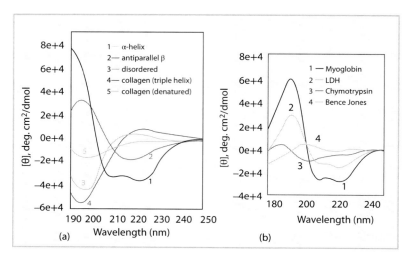

FIGURE 4.8 Circular dichroism (CD) spectra of polypeptides and proteins with representative secondary structures. (a) CD spectra of poly-L-lysine: at pH 11.1 in the α-helical and antiparallel β-sheet conformations; at pH 5.7 in the extended conformations, and the placental collagen in its native triple-helical and denatured forms. (b) CD spectra of representative proteins with varying conformations: sperm whale myoglobin; chicken heart lactate dehydrogenase; bovine α-chymotrypsin; and human Bence Jones protein which is a human immunoglobulin light chain of κ type. (Greenfield, N.J., *Nat. Protoc.*, 1, 2876–2890, 2007.)

4.7 ISOTHERMAL TITRATION CALORIMETRY

Isothermal titration calorimetry (ITC) is a new technique that is utilized for measuring thermodynamic parameters, as well as the equilibrium constant between a ligand and a macromolecule. This technique operates by measuring the heat change during the reaction between two interacting species. It can estimate the binding parameters between two or more reacting species, including two macromolecules. The measuring device has two identical cells that consist of highly conducting materials but that are inert in nature, thus being similar to gold. The reference cell contains only the buffer in which the reaction takes place, the sample cell consists of the macromolecule in the buffer, and both the cells are surrounded by adiabatic jackets. When aliquots of the ligand are added to the sample cell, such as any titration, heat change will take place due to the reaction, exothermic or endothermic in nature, in the

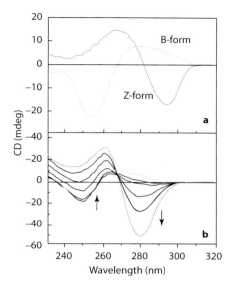

FIGURE 4.9   CD spectra of different DNA. (Scientific Reports 5, Article number: 9943, doi:10.1038/srep09943.)

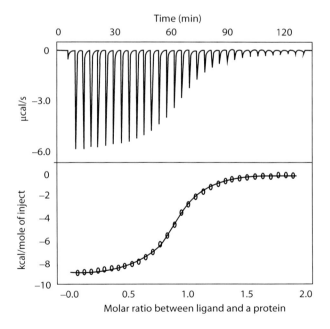

FIGURE 4.10   A typical calorimetric titration of ligand binding to a protein. (http://www.personal.psu.edu.)

opposite sense. The calorimeter keeps the temperature in both reference cells and sample cells the same by varying the power input in a time-dependent manner; the variation in power ultimately measures the heat uptake or heat release. Thus, measurements consist of the time-dependent input of power required to maintain equal temperatures in both sample cells and reference cells. Experiments are represented as a series of spikes (Figure 4.10) that measure variations in power with each injection of a ligand, and the signals can then be normalized to estimate the binding parameters.

CHAPTER 5

Macromolecular Assembly and Recognition with Chemical Entities

Folding or conformation of biomolecules is largely guided by their sequence and function; these also control their recognition feature. Folding has a defined directionality and pattern. However, when molecules have no function, their folding does not follow any defined path. Scientists worldwide are trying to emulate the pattern of protein folding in chemical macromolecules that have no defined function; however, so far, their success has been limited. Thus, the molecular recognition in such macromolecules does not follow any specific rules. In 1955, two years after the discovery of the DNA double helix, Giulio Natta proposed that highly isotactic polypropylene synthesized by the Ziegler–Natta catalyst adopts a helical conformation in the crystalline state. This was the beginning of the concept of synthetic foldamers. *Synthetic foldamers* are defined as polymers that possess the ability to adopt compact conformation. This is a challenge, because unlike proteins, synthetic polymers favor the path of entropy-driven random coil formation (Box 5.1).

BOX 5.1   ZIEGLER–NATTA CATALYST

A Ziegler–Natta catalyst is a titanium-based catalyst named after Karl Ziegler and Giulio Natta. It is used in the synthesis of polymers of alpha-olefins. Natta discovered that when this catalyst is used to polymerize alkenes such as propylenes, it produces crystalline polymers with a stereo regularity. Here, the polymerization is stereospecific and produces linear and not branched polymers.

In the figure accompanying this box, one may notice the regularity of the polymer produced with the Ziegler–Natta catalyst from $MgCl_2$-nEtOH. This crystal structure determined by Ziegler and Natta was responsible for them being awarded the Nobel Prize in 1963. Since its discovery, the catalyst has had tremendous commercial success, especially in producing controlled plastics.

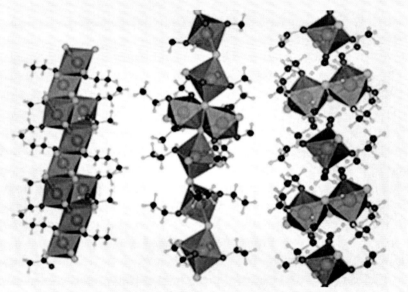

www.chemistryviews.org/details/ezine/1390601/ZieglerNatta_Catalyst_Support_structure.html

The introduction of hydrogen bonding systems in repeating units into polyolefin is an important concept, and the Ziegler–Natta catalyst aided in achieving the same. The nonpolar environment of the polyolefin matrix helps in overcoming the shortcomings of hydrogen bonds and also the entropic barrier that exists with regard to the formation of the H-bonded network that one may encounter during the formation of the assembly.

In the words of Fredga while motivating Ziegler and Natta as they received the Nobel Prize for Chemistry, "Nature synthesizes many stereo regular polymers, for example cellulose and rubber. This ability has so far been thought to be a monopoly, operating with biocatalysts known as enzymes. However, now Professor Natta has broken this monopoly."

5.1 FOLDING AND MOLECULAR RECOGNITION

We have seen in the previous chapters that the folding of a biomacromolecule plays an important role in molecular recognition. In this chapter, we find that the formation of foldamers occurs with specificity in chemical polymers as well, and the Ziegler–Natta catalyst-induced periodicity of polymers is a famous example in that direction. We ask this question: Does this periodicity guide chemical recognition? What are the other means by which this periodicity can be achieved?

5.1.1 Solvent-Induced Folding

In some cases, a nonspecific interaction with solvents leads to conformational ordering in synthetic polymers, as shown in Figure 5.1. This figure

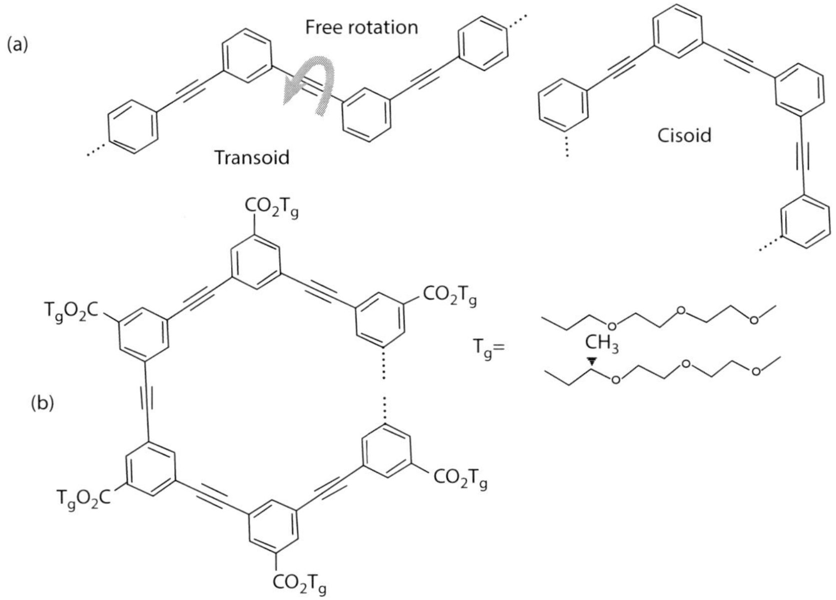

FIGURE 5.1 (a) and (b) Solvent-induced folding of *m*-phenyleneethynylene oligomers developed by Moore and coworkers. (Adapted and modified from Nelson, J.C., Saven, J.G., Moore, J.S., and Wolynes, P.G., *Science*, 277, 1793, 1997.)

shows that in the case of polymers containing an *m*-phenyleneethynylene backbone and triethyleneglycol side chains as hydrophilic segments, *cis* and *trans* conformation takes place along the backbone. Since *cis* and *trans* bonds are positioned at random locations, polymers adopt a random conformation.

5.1.2 Folding due to Charge-Transfer Complex Formation

A conformational degree of freedom of polymers with flexibility is rare due to stiffness of the backbone, bond-angle constraints, and steric factors. Intramolecular charge-transfer complexes between electron-rich donors and electron-deficient acceptors result in flexible polymers, as shown in Figure 5.2.

Here, the charge-transfer complex is formed between dialkoxynaptha-lene, an electron-rich donor, and napthalenetetracarboxylic acid diimide, an electron-deficient acceptor. Both of these are linked in an alternate fashion and are connected by aspartic acid. Oligomers containing both the donor and the acceptor form a periodic charge-transfer complex on mixing.

Later, several such donor–acceptor polymers are designed; they are placed in an alternate fashion and connected by flexible spacers. The key features of these polymers are their ability to form an intramolecular charge-transfer complex, as well as their ability to form a crown ether-type complex with alkali metal ions through molecular recognition (see Chapter 2). The polymers adopt folded conformation in polar solvents as shown in Figure 5.3.

5.1.3 Self-Organization due to Immiscibility

One of the major outcomes of this scheme of organization is the formation of micelles of surfactants in water. This is mainly guided by hydrophobic interactions, as discussed earlier, and follows the same principle as shown in the case of the Langmuir isotherm. Thermodynamic incompatibility segregates the like and unlike molecules, thus promoting self-organization. Surfactants are amphipathic molecules and thus organize themselves between two surfaces: ionic and hydrophobic. Immiscibility with solvents often drives the folding of large-chain molecules.

5.2 HOST–GUEST RECOGNITION AND SUPRAMOLECULAR ASSEMBLY

Macroscopic implications for molecular recognition at the monomeric level are very different, as evident in the case of DNA molecules. However, the recognition principles of both monomers and macromolecules are

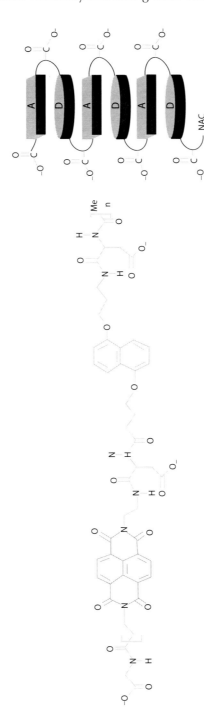

FIGURE 5.2 (See color insert.) Charge transfer–induced folding of a donor–acceptor oligomer. (Adapted and modified from Macmillan Publishers Ltd., Nature, Lokey, R.S., and Iverson, B.L., Synthetic molecules that fold into a pleated secondary structure in solution, 375, 303, copyright 1995.)

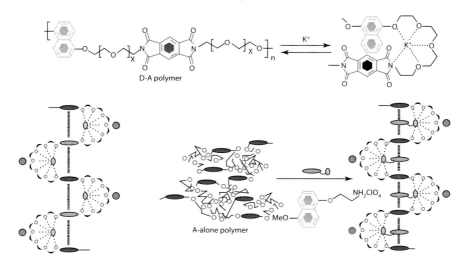

FIGURE 5.3 **(See color insert.)** Charge transfer–induced folding of truly flexible donor–acceptor polymer and acceptor-alone polymer. (Adopted and modified from Ghosh, S., and Ramakrishnan, S., *Angew Chem. Int. Ed.*, 43, 3264, 2004; 44, 5441, 2005.)

more or less the same, although thermodynamic stability is the major determining factor for macromolecular recognition. Here, we focus on a few test cases to depict the assembly of larger aggregates. Well-defined molecular recognition events can be used to direct the assembly of macroscopic objects into larger aggregated structures. Chapter 2 deals with crown ether recognition of alkali metals. These metals are at the border of chemical and biological recognition, as the crown ethers resemble ionophores with important biological functions.

Acrylamide-based gels that were functionalized with either host (cyclodextrin) rings or guest moieties made of hydrocarbons such as adamentyl, butyl groups were synthesized. It was observed that pieces of host and guest molecules adhere together while exhibiting specificity, as shown in Figure 5.4. This adherence takes place through the mutual molecular recognition of the cyclodextrins and hydrocarbon groups on their surfaces. Different gels can be assembled selectively by changing the size and shape of both the host and the guest. Then, they are sorted into distinct macroscopic structures that are in the order of millimeters to centimeters in size.

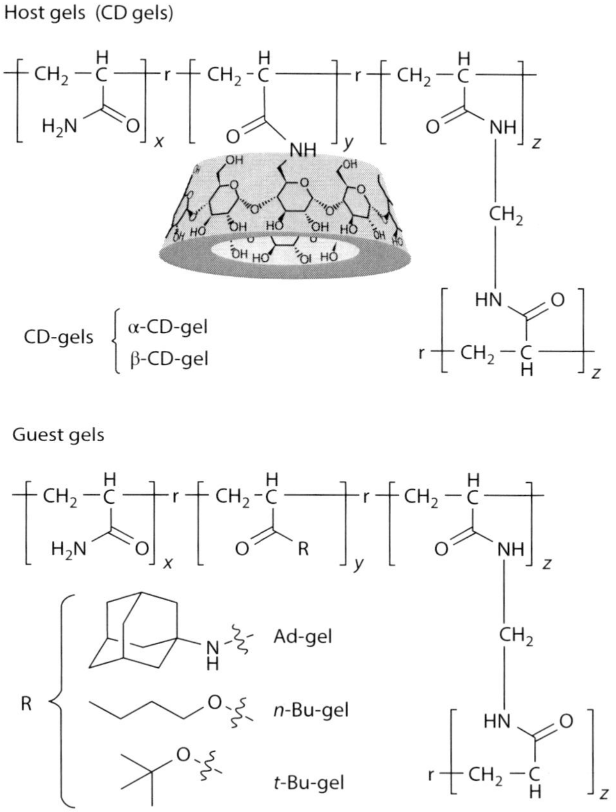

FIGURE 5.4 Chemical structures of host and guest gels. Host gels: α-cyclodextrin-gel and β-CD-gel; guest gels: adamantyl gel (Ad-gel), *n*-butyl-gel (*n*-Bu-gel), and t-butyl-gel (t-Bu-gel). α- and β-CDs are cyclic oligosaccharides consisting of six or seven glucopyranose units, respectively; these are attached by α-1,4-linkages. The character "r" in the main chain of the polymers indicates that each monomer was randomly copolymerized. The molar ratio of acrylamide, host- or guest-modified acrylamide, and N,N'-methylenebis (acrylamide) is shown as *x*, *y*, and *z*. In this study, host and guest gels with a molar ratio *x*:*y*:*z* of 0.948:0.047:0.005 were used. (Harada, A. et al., *Nature Chemistry* 3, 34–37, 2011.)

This interaction can be monitored exclusively by following the principle of molecular recognition.

Figure 5.4 shows the different ratios of the reacting species that were employed to achieve this specific recognition. The strength of the interaction is unique and is shown in Figure 5.5.

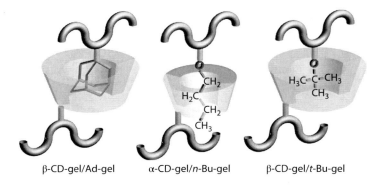

β-CD-gel/Ad-gel   α-CD-gel/*n*-Bu-gel   β-CD-gel/*t*-Bu-gel

FIGURE 5.5 Proposed structures of the complexes formed between cyclodextrin (CD)-gel acting as host and guest molecule made of Adamentyl (Ad) or butyl (Bu) groups with high binding affinity. When the assembly of β-CD-gel with Ad-gel was pulled from both sides, one of the gel pieces broke without damaging the contact interfaces. This indicated that the Ad group on the contact interface of the gels presumably forms an inclusion complex with β-CD in the β-CD-gel. This is done in a similar manner to the formation of complexes between soluble Ad-functionalized polymers and β-CD in homogeneous aqueous solutions. The linear shape of the *n*-Bu group is considered as fitting well into the cavity of α-CD, and the bulky t-Bu group can be incorporated into the larger cavity of β-CD. (Harada, A. et al., *Nature Chemistry* 3, 34–37, 2011.)

5.3 MOLECULAR INFORMATION PROCESSING AND SELF-ORGANIZATION

According to Jean-Marie Lehn, "The selective binding of a substrate by a molecular receptor to form a supramolecular species involves molecular recognition which rests on the molecular information stored in the interacting species."

If a device is considered a tool for information processing and signal generation, then supramolecular association, organization, and function present a great future. The inherent plasticity both in the molecule and at the recognition surface results in diversity of an unparalleled dimension. We have seen the abundance of information processing in biological assembly. Several devices can be developed based on the principle of molecular recognition. The components of these devices can be photoactive, thermosensitive, chemoactive, or ionophoric in nature. Two concepts should be clear before initiating the design of such materials: the materials should be able to perform a given function and capable of forming an organized array.

In 1987, the Nobel Prize for Chemistry was awarded to Donald J. Cram, Jean-Marie Lehn, and Charles J. Pedersen for the development exhibited in the area of supramolecular chemistry. In this area, they demonstrated the use of cryptands and crown ethers and launched a new area of host–guest relationship.

Cryptands are a family of synthetic bi- and polycyclic multidentate ligands that form complexes with cations like variety of alkali metal ions. "The term cryptand implies that this ligand binds substrates in a crypt, interring the guest as in a burial." They are usually more selective than ionophores, lipophilic, and three-dimensional in nature (Figure 5.6).

Cryptands exhibit better binding strength and selectivity with cations than with ionophores and can bind metal ions in organic solvents. They are also used as phase-transfer catalysts.

Electroactive devices are synthesized from organic assembly, and they can transmit information and conduct ionic signals. Polyolefinic chains, as prepared by the Ziegler–Natta catalyst, can act as molecular wires for

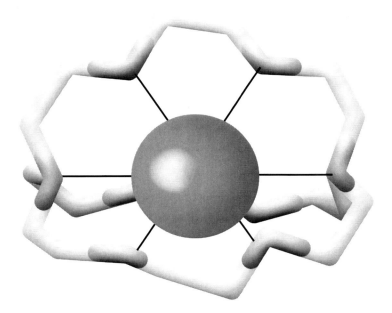

FIGURE 5.6 Structure of 2.2.2-cryptand encapsulating a potassium cation (purple)—at a crystalline state, obtained with x-ray diffraction. (Adapted and modified from https://upload.wikimedia.org/wikipedia/commons/thumb/9/98/ Cryptate_of_pottasium_cation.jpg/220px-Cryptate_of_pottasium_cation.jpg.)

electron transport across membranes. According to Lehn, polyolefins exhibit very good optical properties.

Macrocyclic compounds can also be used at the air–water interface to generate the Langmuir–Blodgett film. However, in many cases, supramolecular chemistry relies on a preorganized structure for generating molecular recognition. The challenge that chemists face would be designing molecules that would be organized by self-motivation that is guided entirely by the structure. Designing a system that is capable of self-organization such as DNA or proteins is indeed a difficult task. The inherent directionality in biopolymers makes one think that this may serve as another important parameter for self-organization. Metal ion–induced organization and periodicity in chemical polymers offers a big hope. Once a method is standardized enough to synthesize self-organized macromolecules, newer properties with a variety of functions will emerge.

5.4 RECOGNITION BETWEEN METAL IONS AND NUCLEIC ACIDS

In Chapter 3 (Box 3.2), we discussed the use of cisplatin as an anticancer drug. Several similar metal-based interactions between nucleic acids and proteins play a pivotal role in biology. The charge-based interactions between polyanion-like nucleic acids and positively charged metal ions have several levels of diversity; these levels depend on both the oxidation states of metal ions and their electronic configuration.

The entire range of this study deals with the interactions between metal ions and nucleic acids that evolved mainly due to three reasons: (1) participation of metal ions in almost all biological processes involving nucleic acids; (2) use of metal ions in some important physicochemical studies— inducing stability/destability of helix, separation of DNA and RNA by the depolymerization technique, and (3) separation of satellite DNA by a base-specific reaction. However, various effects of metal ions on the structure and function of nucleic acid can be best studied by knowing the structure and stability of the metal complexes that are formed with the monomeric units of nucleic acids: nucleotides and nucleosides. A nucleotide, consisting of a heterocyclic base, phosphate, and a sugar unit, offers itself as a potential donor source. Since a metal can bind to the phosphate portion, the base, or the hydroxyl of the ribose sugar and because each base offers a number of sites, the total number of sites available for metal binding would be quite large. Therefore, an inorganic chemist would be interested in checking out the location of the metal ion when such a situation arises.

The replication of DNA undergoes misincorporation when Mg(II) is replaced by Mn(II) in the presence of DNA polymerase. Actually, the activation of the DNA polymerase enzyme requires a bivalent metal, along with the triphosphate precursors of nucleic acids, such as dATP, dGTP, dTTP, and dCTP. The divalent metal that is generally used is Mg(II), and the reaction involves the cleavage of pyrophosphate from the triphosphate and the subsequent formation of the phosphodiester bond between two adjacent deoxynucleotides. It was earlier reported that whenever Mn(II) is used instead of Mg(II), the selectivity for deoxynucleotides is greatly reduced and ribonucleotides can be randomly incorporated. Later, Khorana and others performed an elaborate study on the kinetics of the incorporation of deoxynucleotides and ribonucleotides at different temperatures and found that at 37°C, the rate is 1:1.

The transcription of the DNA code to mRNA is brought about by RNA polymerase enzymes; again, this process takes place in the presence of metal ions. These polymerases require DNA as a template and they also need triphosphates of ribonucleotides. Several metal ions such as Mg(II), Mn(II), and Co(II) have been shown to be active; with RNA polymerase, even Mg(II) exhibits a high selectivity for the substrate. The presence of Mg(II) ensures accurate incorporation of ribonucleotides instead of deoxynucleotides; whereas while quantitatively promoting more rapid reactions, Mn(II) can induce errors by incorporating deoxyucleotides along with ribonucleotides. This difference in the behaviors of Mg(II) and Mn(II) for both replication and transcription mechanisms has been attributed to the varying coordination tendencies. Other metal ions such as Cu(II), Cd(II), and Hg(II) are effective inhibitors of the Mg(II)- and Mn(II)-activated mammalian RNA polymerase. The final step of protein synthesis from its constituent amino acids is governed by the base sequence of mRNA and is also very much dependent on the metal ion concentration. Generally, in this case, even Mg(II) is required for *in vitro* protein synthesis. The misincorporation of amino acids in the polypeptide chain has been observed to take place when the concentration of Mg(II) is varied. Gunther Eichhorn proposed that at a high Mg(II) concentration, all the three letters of the genetic code can be misread leading to erroneous incorporation of aminoacids during translation.

5.4.1 Stabilization of Nucleic Acids by Metal Ions

It was initially observed that double-helical DNA denatures in distilled water in the absence of electrolyte in the mM concentration range. The unwinding of the strands then takes place mainly due to the charge repulsion of adjacent phosphate groups. Metal ions with counter charges can stabilize the double helix in solution. Many factors contribute to the stability of the ordered polynucleotide structures. There are hydrophobic forces that bring two bases close enough to form a hydrogen bond and that also allow them to stack over each other. In addition, counter ions interact electrostatically with the negatively charged phosphate, thereby minimizing any charge repulsion and subsequent unwinding of the helix. It was observed that T_m, the melting temperature or the temperature at which DNA unwinds into single strands, increases linearly with the logarithm of ionic strength. Divalent ions [Mg(II) and Co(II)] are effective at much lower concentrations than are univalent ions. It has also been found that the addition of reducing agents, such as ascorbic acid to the DNA–metal complex, results in an increase in the T_m of DNA. This seems to indicate that a change in the oxidation state of the metal ion results in a change in the stability of the DNA helix. However, this stabilization of the double helix is not a universal characteristic of metal ions.

In fact, different metals have different effects on the conformation as well as on the stability of DNA. The following three broadly distinct categories of metal ions could be considered: (i) those that act as counter ions and stabilize the helix by electrostatic interactions with the phosphate group, such as alkali and alkaline earth metal ions; (ii) those that destabilize the helix by coordinating through N-containing bases, such as Hg(II), Ag(I), and Pt(II); and (iii) those that act on phosphates and bases simultaneously, such as Cu(II), Fe(II), Co(II), and Mn(II). In addition, a few other metal ions, such as Co(III) and B(III), also interact with the ribose moiety. The stabilizing and destabilizing effects of these metals on the DNA helix were initially clearly illustrated in the case of an interaction study that was carried out between Mg(II) and Cu(II) with DNA. If we consider that the double helix has a low absorbance value in comparison with a random coil structure, heat denaturation will cause a transition from low absorbance to high absorbance. Subsequent cooling shows a decrease in absorbance resulting from H-bonding within the single strands, thereby forming hairpin loops.

The randomly oriented single chains cannot reunite on cooling, as this is energetically unfavorable. This melting transition of DNA was found to occur at a higher temperature when Mg(II) was used as the counter ion. This is expected, because stabilization of the helix by bivalent metal-like Mg(II) always results in an unwinding that is associated with high energy. When Cu(II) was used instead of Mg(II), a contrary phenomenon was observed in the melting of DNA. This is probably due to the fact that Cu(II) interacts with bases of DNA, thus reducing the possibility of it forming an H-bond between complementary bases. As a result of this, the helix could probably collapse at a lower temperature. The cooling of Cu(II)–DNA solution is not accompanied by a decrease in the absorbance to form hairpins. It is explained on the basis that Cu(II) still binds to the bases of DNA and prevents any H-bond formation in the same strand. If a high ionic strength is maintained in the medium, Cu(II) fails to bring about destabilization of the helix. The rewinding of the two DNA strands could be achieved by either the addition of Na(I) ion or the removal of Cu(II) ion, by either a complexing agent, such as EDTA, or dialysis.

Cu(II) ions are not unique in their ability to bring about reversible unwinding and rewinding of DNA. In fact, Zn(II) can be made to induce reversible unwinding just by manipulating temperature. This apparently different behavior of Zn(II) from that of Cu(II) may be due to the fact that Zn(II) binds less strongly with DNA bases than does Cu(II). It has been pointed out that many metal ions bind to both phosphate and base sites, and their effects on DNA depend on their relative affinity for the two types of binding sites. Mn(II), Zn(II), Cd(II), and Cu(II) exhibit an increase in T_m at a lower metal ion concentration; however, after a certain maximum concentration is reached, the reverse trend is noticed. Obviously, all these metals stabilize DNA at lower concentrations by phosphate binding. This stabilizing effect is counteracted by base binding of the metal at a higher concentration, leading to the rupture of H-bonds. The affinity ratio of metal ions for binding to both phosphates and DNA base sites generally follows the following order: Mg(II) > Co(II) > Ni(II) > Mn(II) > Zn(II) > Cd(II) > Ca(II). This concept has profound influence on any biological recognition process in which metal ions participate.

It had been proposed earlier that the conformational changes in DNA during the recognition process caused by metal ions were very much dependent on the ionic strength of the medium. The change in conformation disappears in the concentration range of 0.5–1.0 M Na(I) and reappears again at a higher ionic strength. These structural alterations probably result from both the shielding of negatively charged phosphate groups and the breakdown of the water structure along the DNA helix. The metal-induced conformational changes of DNA could be explained on the basis of alteration of the winding angle between base pairs that occurs in the transition from B to other conformation of DNA.

5.4.2 Metal Ion Interaction with Nucleosides and Nucleotides

It should be emphasized here that the participation of metal ions in macromolecular recognition is primarily based on charge; this is accompanied by the participation of the water shell around these ions. Many crystal structures reveal that protonation or deprotonation of water by metal ions facilitates several biochemical reactions. Table 5.1 provides a comprehensive account of all pK_a values of nucleic acid constituents; these values ultimately guide the charge-based recognition (for numbering, refer to Figures 1.3 and 1.4, Chapter 1).

During the recognition process, various metal ions perform different functions depending on their size and mobility. Where mobility is at a premium in signal transduction, energy transfer, and so on, clearly alkali metals or alkaline earth metals such as Na(I), K(I), Mg(II), and Ca(II) will be preferred due to their small size and electronic configuration. On the other hand, larger cations such as Mn(II), Fe(III), Cu(II), Ni(II), and Mo(III) participate in redox chemistry without much mobility. Due to vacant d-orbitals, many of the metals in the class of transition metals take part in electron-transfer recognition. Relatively weak and inert metals such as Zn(II) or Cd(II) are useful for acid–base reactions.

During enzyme-based biochemical recognition, in all the major classes of catalytic reactions, such as electron-transfer reactions, acid catalysis,

TABLE 5.1 pK_a Values of Different Nucleosides and Nucleotides

Nucleosides/Nucleotides	Probable Ionization Sites	pK_a Values
Adenosine	N1, ribose	3.6 (basic), 12.5 (acidic)
Guanosine	N7/C6, ribose	1.8 (basic), 9.5 (acidic)
Cytidine	N1, ribose	4.2 (basic), 12.4 (acidic)
Thymidine	N1-H/C6, ribose	9.2 (basic), 12.5 (acidic)

Note: In the case of nucleotides, the phosphate ionization constant lies between 5·8 and 6·6.

atom transfer, or free radical reactions, specific metal ions are used as per the specificity of the reaction. For example, paramagnetic transitional metals are required for free radical reactions. Similarly, for Lewis acid–catalyzed reactions, one needs an open coordination site, easily replaceable ligands, and functions that could be fulfilled by Zn(II).

The title of this chapter suggested that recognition principles between chemical entities would be discussed. Although DNA is a master biological molecule, in this chapter we have considered it a chemical macromolecule and also studied its interaction pattern with metal salts. Throughout this book, DNA is mentioned repeatedly in terms of its structure and various functions, which enable the molecule to stand tall among others. A chemist's dream would be to discover a new molecule, such as DNA, that would define the paradigm of molecular recognition.

Suggested Readings

CHAPTER 1

Berg, J.M., Tymoczko, J.L., and Stryer, L. (2002) *Biochemistry*, 5th edition, New York: W H Freeman.

Branden, C., and Tooze, J. (1999) *Introduction to Protein Structure*, 2nd edition, Garland Science: Taylor & Francis Group.

Chakrabarti, K.S., Agafonov, R.V., Pontiggia, F., Otten, R., Higgins, M.K., Schertler, G.F., Oprian, D.D., Kern, D. (2016) Conformational Selection in a Protein-Protein Interaction Revealed by Dynamic Pathway Analysis. *Cell Rep*, 14, 32.

Eliel, E.L. (1975) *Stereochemistry of Carbon Compounds*, New Delhi: Tata McGraw-Hill.

Koshland, D.E. Jr. (1958) Application of a theory of enzyme specificity to protein synthesis, *Proc Natl Acad Sci*. 44, 98.

Kuriyan, J., Konforti, B., and Wemmer, D. (2012) *The Molecules of Life*, 1st edition, Garland Science: Taylor & Francis Group.

Newcomer, M.E., Lewis, B.A., and Quiocho, F.A. (1981) The radius of gyration of L-arabinose-binding protein decreases upon binding of ligand, *J Biol Chem*. 256, 13213.

CHAPTER 2

Bharati, B.K., and Chatterji, D. (2013) Quorum sensing and pathogenesis: Role of small signaling molecules in bacterial persistence, *Curr Sci*. 105, 643.

Jacob, F., and Monod, J. (1961) Genetic regulatory mechanisms in the synthesis of proteins, *J Mol Biol*. 3, 318.

Knox, J.R., and Pratt, R.F. (1990) Different modes of vancomycin and D-alanyl-D-alanine peptidase binding to cell wall peptide and a possible role for the vancomycin resistance protein, *Antimicrob Agents Chemother*. 34, 1342.

Lewin, B., Krebs, J., Kilpatric, S.T., and Goldstein, E.S. (2011) *Lewin's Genes X*, Sudbury, MA: Jones and Bartlett Learning.

Pesavento, C., and Hengge, R. (2009) Bacterial nucleotide-based second messengers, *Curr Opin Microbiol*. 12, 170.

Raffa, R.B., and Porreca, F. (1989) Thermodynamic analysis of the drug-receptor interaction, *Life Sci.* 44, 245.

Romling, U., Galperin, M.Y., and Gomelsky, M. (2013) Cyclic di-GMP: The first 25 years of a universal bacterial second messenger, *Microbiol Mol Biol Rev.* 77, 1.

Shanahan, C.A., and Strobel, S.A. (2012) Quorum sensing and pathogenesis: Role of small signaling molecules in bacterial persistence, *Org Biomol Chem.* 10, 9113.

Steed, J.W. (2001) First- and second-sphere coordination chemistry of alkali metal crown ether complexes, *Coord Chem Rev.* 215, 171.

Thamotharan, S., Karthikeyan, T., Kulkarni, K.A., Shetty, K.N., Surolia, A., Vijayan, M., and Suguna, K. (2011) Modification of the sugar specificity of a plant lectin: Structural studies on a point mutant of *Erythrina corallodendron* lectin, *Acta Crystallogr D Biol Crystallogr.* 67, 218.

Watson, J.D., and Crick, F.H.C. (1953) Molecular structure of nucleic acids, *Nature.* 171, 737.

CHAPTER 3

Anfinsen, C.B., and Haber, E. (1961) Studies on the reduction and re-formation of protein disulfide bonds, *J Biol Chem.* 236, 1361.

Anfinsen, C.B., Haber, E., Sela, M., and White, F.H., Jr. (1961) The kinetics of formation of native ribonuclease during oxidation of the reduced polypeptide chain, *Proc Natl Acad Sci.* 47, 1309.

Arnold, A.R., and Barton, J.K. (2013) DNA protection by the bacterial ferritin Dps via DNA charge transport, *J Am Chem Soc. 135*, 15726.

Bielawski, C., and Chen, Y.S. (1998) A modular approach to constructing multisite receptors for isophthalic acid, *Chem Commun.* 12, 1313.

Branden, C.I., and Tooze, J. (1999) *Introduction to Protein Structure*, 2nd edition, New York: Garland Science.

Breiten, B., Mathew, R.L., Sherman, W. et al. (2013) Water networks contribute to enthalpy/entropy compensation in protein–ligand binding, *J Am Chem Soc.* 135, 15579.

Busby, S., and Ebright, R. (1999) Transcription activation by catabolite activator protein (CAP), *J Mol Biol.* 293, 199.

Calladine, C.R., and Drew, H.R. (1997) *Understanding DNA: The Molecule and How It Works*, 2nd edition, San Diego, CA: Academic Press.

Carlos, C.E. et al., (2011) *Nature Methods*, 8, 3.

Castro, C.E., Kilchherr, F., Kim, D.N., Lin Shiao, E., Wauer, T., Wortmann, P., Bathe, M. and Dietz, H. (2011) A primer to scaffolded DNA origami, *Nat Methods*, 8, 221.

Chatterji, D., and Gopal, V. (1996) Fluorescence spectroscopy analysis of active and regulatory sites of RNA polymerase, *Methods Enzymol.* 274, 456.

Dean, L. (2005) *Blood Groups and Red Cell Antigens*, Bethesda, MD: National Center for Biotechnology Information (US).

DiMaio, F., Yu, X., Rensen, E., Krupovic, M., Prangishvili, D., and Egelman, E.H. (2015) A virus that infects a hyperthermophile encapsidates A form DNA, *Science*. 348, 914.

Drlica, K., and Rouviere-Yaniv, J. (1987) Histonelike proteins of bacteria, *Microbiol Rev.* 51, 301.

Friedman, D.I. (1988) Integration host factor: A protein for all reasons, *Cell.* 55, 545.

Gellman, S.H. (1997) Introduction: Molecular recognition, *Chemical Rev.* 97, 1231.

Gerling, T., Wagenbauer, K.F., Neuner, A.M., and Dietz, H. (2015) Dynamic DNA devices and assemblies formed by shape-complementary, non–base pairing 3D components, *Science*. 347, 1446.

Ghatak, P., Karmakar, K., Kasetty, S., and Chatterji, D. (2011) Unveiling the role of DPS in the organization of mycobacterial nucleoid, *PLoS ONE.* 6, e16019.

Gupta, S., and Chatterji, D. (2003) Bimodal protection of DNA by *M. smegmatis* DNA binding protein from stationary phase cells, *J Biol Chem.* 278, 5235.

Ha, S.C., Lowenhaupt, K., Rich, A., Kim, Y.G., and Kim, K.K. (2005) Crystal structure of a junction between B-DNA and Z-DNA reveals two extruded bases, *Nature.* 437, 1183.

Jones, S., Barker, J.A., Nobeli, I., and Thornton, J.M. (2003) Using structural motif templates to identify proteins with DNA binding function, *Nucleic Acids Res.* 31, 2811.

Kimsey, I.J., Petzold, K., Sathyamoorthy, B., Stein, Z.W., and Al-Hashimi, H.M. (2015) Visualizing transient Watson–Crick-like mispairs in DNA and RNA duplexes, *Nature.* 519, 315.

Kuby, J., Goldsby, R.A., Kindt, T.J., and Osborne, B.A. (2000) *Immunology*, 4th edition, Gordonsville, VA: W H Freeman & Co.

Kuriyan, J., Konforti, B., and Wemmer, D. (2012) *The Molecules of Life*, 1st edition, New York: Garland Science, Taylor & Francis Group.

Lee, R.C., Feinbaum, R.L., and Ambros, V. (1993) The *C. elegan* sheterochronic gene lin-4 encodes small RNAs with antisense complementarity to lin-14, *Cell.* 75, 843.

Lehn, J.M. (1995) *Supramolecular Chemistry: Concepts and Perspectives*, Weinheim: Wiley VCH.

Lewin, B., Krebs, J., Kilpatric, S.T., and Goldstein, E.S. (2011) *Lewin's Genes X*, Sudbury, MA: Jones and Bartlett Learning.

Lodish, H., Berk, A., Zipursky, S.L. et al., (2000) *Molecular Cell Biology*, 4th edition, New York: W H Freeman and Company.

MacDonald, D., Demarre, G., Bouvier, M., Mazel, D., and Gopaul, D.N. (2006) Structural basis for broad DNA-specificity in integron recombination, *Nature.* 440, 1157.

Murakami, K.S., and Darst, S.A. (2003) The bacterial RNA polymerases: The whole story, *Curr Opin Struc Biol.* 13, 31.

Nakamura, T., Zhao, Y., Yamagata, Y., Hua, Y.J., and Yang, W. (2012) Watching DNA polymerase η make a phosphodiester bond, *Nature*. 487, 196.

Nelson, D.L., and Cox, M.M. (2005) *Lehninger Principles of Biochemistry*, 4th edition, New York: W H Freeman and Company.

Ptashne, M. (2004) *The Genetic Switch*, 3rd edition, New York: Cold Spring Harbor.

Ptashne, M., and Gann, A. (2002) *Genes and Signals*, New York: Cold Spring Harbor.

Ramakrishnan, V. (2002) Ribosome structure and the mechanism of translation, *Cell*. 108, 557.

Rosenberg, B., VanCamp, L., Trosko, J.E., and Mansour, V.H. (1969) Platinum compounds: A new class of potent antitumour agents, *Nature*. 222, 385.

Rothemund, P.W. (2006) Folding DNA to create nanoscale shapes and patterns. *Nature*. 440, 297.

Saenger, W. (1984) *Principles of Nucleic Acid Structure*, New York: Springer-Verlag.

Sarkar, P., Sardesai, A.A., Murakami, K.S., and Chatterji, D. (2013) Inactivation of the bacterial RNA polymerase due to acquisition of secondary structure by the omega subunit, *J Biol Chem*. 288, 25076.

Sharma, U.K., and Chatterji, D. (2010) Transcriptional switching in *E. coli* during stress and starvation by modulation of sigma - 70 activity, *FEMS Microbiol Rev*. 34, 647.

Storry, J.R., and Olsson, M.L. (2009) The ABO blood group system revisited: A review and update, *Immunohematology*. 25, 48.

Sunderberg, E.J., and Mariuzza, R.A. (2003) Molecular recognition in antigen antibody complexes, *Advanc Protein Biochem*. 61, 119.

CHAPTER 4

Baranovsky, S.F., Bolotin, P.A., Evstigneev, M.P., and Chernyshev, D.N. (2009) Intercalation of ethidium bromide and caffeine with DNA in aqueous solution, *J Appl Spectrosc*. 76, 132.

Cantor, C.R., and Schimmel, P.R. (1980) *Biophysical Chemistry*, Volumes I and II, San Francisco, CA: W H Freeman.

Ganguly, A., and Chatterji, D. (2011) Sequential assembly of an active RNA polymerase molecule at the air-water interface, *Langmuir*. 27, 3808.

Ganguly, A., and Chatterji, D. (2012) The σ-competition model in *Escherichia coli*: A comparative kinetic and thermodynamic perspective, *Biophys J*. 103, 1325.

Ganguly, A., Rajdev, P., Williams, S.M., and Chatterji, D. (2012) Non-specific interaction between DNA and protein allows for cooperativity: A case study with mycobacterial DNA binding protein, *J Phys Chem B*. 116, 621.

Greenfield, N.J. (2006) Using circular dichroism spectra to estimate protein secondary structure, *Nat Protoc*. 1, 2876.

Hellman, L.M., and Fried, M.G. (2007) Electrophoretic mobility shift assay (EMSA) for detecting protein–nucleic acid interactions, *Nat Protoc*. 2, 1849.

Lakowicz, J.R. (2006) *Principles of Fluorescence Spectroscopy*, 3rd edition, New York: Springer.

Melhuish, W. H. (1964) Measurement of quantum efficiencies of fluorescence and phosphorescence and some suggested luminescence standards. *J. Opt. Soc. Am.* 54, 183.

Minasyants, M.V. (2014) Investigation of differential absorption of DNA complexes with ligands, *Chem Biol.* 1, 51.

Pierce, M.M., Raman, C.S., and Nall, B.T. (1999) Isothermal titration calorimetry of protein protein interactions, *Methods.* 19, 213.

Porath, J., and Olin, B. (1983) Immobilized metal affinity adsorption and immobilized metal affinity chromatography of biomaterials. Serum protein affinities for gel-immobilized iron and nickel ions, *Biochemistry.* 22, 1621.

Schasfoort, R.B.M., and Anna, J.T. (2008) *Handbook of Surface Plasmon Resonance,* Cambridge: RSC Publishing.

CHAPTER 5

Ghosh, S., and Ramakrishnan, S. (2004) Aromatic donor-acceptor charge-transfer and metal-ion complexation assisted folding of a synthetic polymer, *Angew Chem Int Ed.* 43, 3264.

Ghosh, S., and Ramakrishnan, S. (2005) Small-molecule-induced folding of a synthetic polymer, *Angew Chem Int Ed.* 44, 5441.

Harada, A., Kobayashi, R., Takashima, Y., Hashidzume, A., and Yamaguchi, H. (2011) Macroscopic self-assembly through molecular recognition, *Nat Chem.* 3, 34.

Hoff, R., and Mathers, R.T. (2010) *Handbook of Transition Metal Polymerization Catalysts,* Hoboken, NJ: John Wiley & Sons.

Lehn, J.M. (1990) Perspectives in supramolecular chemistry—From molecular recognition towards molecular information processing and self-organization, *Angew Chem Int Ed.* 29, 1304.

Lehn, J.M. (1995) *Supramolecular Chemistry: A Personal Account Concepts and Perspectives,* 1st edition, Weinheim: Wiley VCH.

Lokey, R.S., and Iverson, B.L. (1995) Synthetic molecules that fold into a pleated secondary structure in solution, *Nature.* 375, 303.

Malizia, F., Fait, A., and Cruciani, G. (2011) Crystal structures of Ziegler-Natta catalyst supports, *Chemistry.* 17, 13892.

Nelson, J.C., Saven, J.G., Moore, J.S., and Wolynes, G. (1997) Solvophobically driven folding of nonbiological oligomers, *Science.* 277, 1793.

Index